高等院校土建学科双语教材（中英文对照）

◆ 给水排水工程专业 ◆

BASICS

水循环系统

WATER CYCLES

［德］桃丽丝·哈斯－阿尔恩特　编著

柳美玉　杨　璐　译

中国建筑工业出版社

著作权合同登记图字：01-2009-7706号

图书在版编目（CIP）数据

水循环系统/（德）哈斯-阿尔恩特编著；柳美玉，杨璐译．—北京：中国建筑工业出版社，2011.2

高等院校土建学科双语教材（中英文对照）◆给水排水工程专业◆

ISBN 978-7-112-12866-2

Ⅰ.①水… Ⅱ.①哈…②柳…③杨… Ⅲ.①水循环-汉、英 Ⅳ.①P339

中国版本图书馆CIP数据核字（2011）第007555号

责任编辑：孙 炼
责任设计：陈 旭
责任校对：陈晶晶 马 赛

高等院校土建学科双语教材（中英文对照）
◆给水排水工程专业◆
水循环系统
［德］桃丽丝·哈斯-阿尔恩特 编著
柳美玉 杨 璐 译

*

中国建筑工业出版社出版、发行（北京西郊百万庄）
各地新华书店、建筑书店经销
北京嘉泰利德公司制版
北京云浩印刷有限责任公司印刷

*

开本：880×1230毫米 1/32 印张：4⅞ 字数：153千字
2011年5月第一版 2011年5月第一次印刷
定价：**18.00**元
ISBN 978-7-112-12866-2
（20133）

中文部分目录

CONTENTS

序

对于发达的工业化国家，饮用水*是一个建筑所必须具备的基本条件。然而，饮用水是一种非常重要的日用品，而且在世界的许多地方非常稀缺。不仅对于工业化国家，对于消费者而言，饮用水的收集和处理成本越来越高。同时，由于废水中所含的污染物质，废水的排放和净化越来越复杂，同时也需要消耗越来越多的资源。

对于建筑师而言，饮用水和废水之间存在一点共同之处——在进行建筑设计的时候，非常重要的一项内容就是饮用水和废水的布置、使用以及排放。饮用水和废水管道系统的布置以及相关的技术需求将对厨房、卫生间的区域造成一定的影响。避免饮用水的过多消耗是在进行建筑设施设计时的一项重要内容。

在对一栋建筑进行设计的时候，必须对如何减少水的消耗的可能性和要求等问题有一个根本性的了解，其中最重要的是要对不同因素之间的相互联系和影响以及相应的技术问题进行了解。在设计时，需要把建筑中的水循环考虑到整体设计中。

本书主要是针对建筑学专业的学生以及没有建筑设备相关知识背景的毕业生而编写出版的。书中还附有一些易懂的说明和解释，能够帮助读者逐步理解本书内容。本书对建筑中不同部位的水管道进行了描述，同时还对它们的功能和技术需求做了讲解，并帮助读者在实际的设计过程中更好地理解它们之间的相关关系。

编辑　伯特·比勒费尔德（Bert Bielefeld）

---

* 原书为 drinking water，但我国所称饮用水是指自来水经过深度处理后的直饮水。而自来水厂供给的冷水，虽然符合饮用水标准，但一般称给水，不叫饮用水。本书保留原文的直译，称为饮用水。——编者注

FOREWORD

The availability of drinking water in buildings is taken for granted in developed, industrialized countries. Drinking water is, however, a valuable commodity, and is scarce in many parts of the world. The collection and treatment of drinking water are becoming increasingly expensive for industrialized countries and therefore also for the consumer. Similarly, the disposal and cleaning of waste water are becoming more resource-intensive and complex due to the substances it contains.

The interface between the drinking water and waste water is the distribution, use and disposal of water within buildings, a significant component of the architect's design. The arrangement of supply and disposal pipework and the technical requirements influence the location of sanitary and kitchen areas. Avoiding high water consumption is an important aspect of technical building services planning.

A broad knowledge of the requirements and possibilities for reducing water consumption is necessary to be able to take these key topics into account in the design of a building, right from its inception. This includes, above all, an understanding of the interrelationships and dependencies, as well as technical systems. It is important to think of the water cycle in a building as an integral part of the design.

The volume *Basics Water Cycles* is aimed at students of architecture and recent graduates without previous knowledge of building services. With the aid of easy-to-understand introductions and explanations, the reader is taken through the subject matter step by step. The path of water through the various zones of a building is described and related to their specific roles and requirements, so that students are able to fully understand the interrelationships and introduce them into their own designs.

Bert Bielefeld, Editor

INTRODUCTION

Part of the technical services in a modern building is a complex pipework system for supplying drinking water and disposing of waste water. This system is a cycle, somewhat similar to the natural water cycle: fresh water is collected, supplied to the building, distributed through a pipework system, and heated if required. It is piped to the draw-off points in bathrooms, kitchens and other sanitary rooms. As soon as it leaves the drinking water pipe through the faucet, it becomes waste water and flows through the waste water pipework into the sewers, from where it is cleaned again and finally returned to natural watercourses. Architects must integrate this cycle into the design of their buildings, as without a carefully planned and properly functioning fresh and waste water system, WCs cannot be flushed, washing machines cannot be operated, and no water will emerge from a shower.

The chapters that follow consider the individual positions of water in a building along the water cycle, and describe the functions of the elements connected to this cycle. It should become clear how a drinking water supply system works, how it is designed into a building, and which aspects should be taken into account. There is also an explanation of how waste water is created and conducted into the drainage system, the general problems that arise in the supply and disposal of water, and the options for their solution.

Approximately two thirds of the earth's surface is covered with water. Of this, only 0.3% is fresh water and therefore potential drinking water. Drinking water is very high-quality fresh water that is suitable for human consumption.

THE NATURAL WATER CYCLE

The natural water cycle—or hydrologic cycle—is a continuous sequence of evaporation, precipitation, and rainwater draining into bodies of open water or seeping into the ground to accumulate as groundwater. Water vapor rises under the influence of solar radiation or other heating effects to form clouds, and falls as precipitation back onto the earth's surface. Some of the rainwater that seeps away is absorbed by the ground, some evaporates, and some is taken up into plants by capillary action. A proportion reaches the lower soil strata and helps maintain the groundwater table. › Fig. 1

Groundwater

Groundwater is described as precipitation water that is stored on top of an impervious stratum and has a temperature of between 8 and 10 °C all year round. Groundwater is generally microbe-free and is pumped up to the surface from deep wells. It provides about three quarters of our drinking water, and goes through several stages of cleaning and filtering before it is fed into the public supply network.

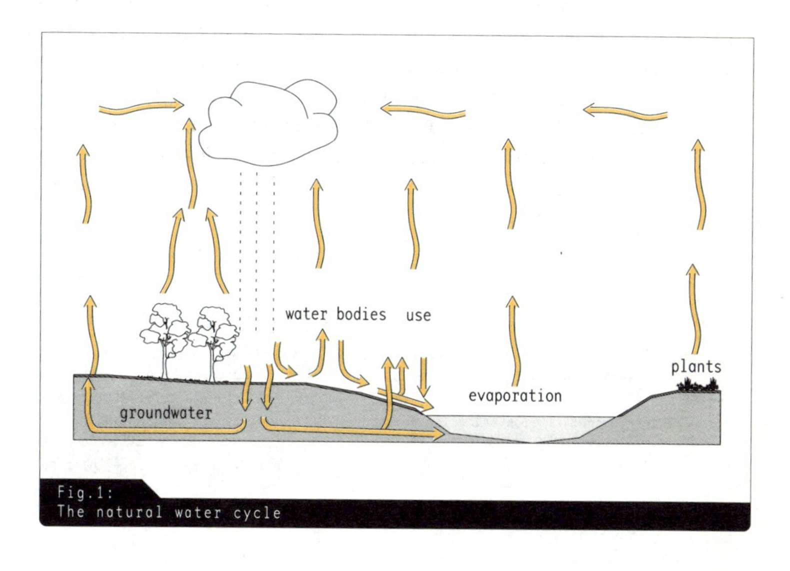

Fig. 1:
The natural water cycle

Substantial groundwater extraction and extensive sealing of urban land surfaces have a substantial impact on the natural water cycle. Rainwater falling on impervious areas cannot seep naturally into groundwater, but is conducted directly into bodies of open water or into the drains. The groundwater table is greatly reduced by building developments, deforestation and drainage works.

In addition, the extensive extraction of groundwater for agriculture and industry, and the pollutants that these activities introduce, are harmful to the system. Pollutants from sources such as manure, agricultural pesticides, landfill, highway drainage and industrial emissions, which fall as acid rain and seep into the groundwater, are a serious cause for concern and can be removed only by expensive cleaning and filtering. The increasing contamination of water combined with high water usage produces an ecological imbalance, the consequences of which result in high costs.

STANDARDS FOR DRINKING WATER

Drinking water intended for human consumption has to meet certain standards. It must be good to taste, odorless and colorless, and free of pathogens and microbes. Every draw-off point must provide best-quality drinking water at sufficient pressure. Quantities of chemicals added to disinfect the water, and of other possible constituents, must be kept within limits specified for European Union countries by an EU directive and regional drinking water regulations. The water's quality and the limits for the substances it contains are checked regularly in accordance with the applicable national standards for drinking water. These standards for drinking water quality change constantly. Today's level of pollution means they can be met only with great difficulty and at increasing cost.

Hardness level

Water with high calcium and magnesium content is described as hard, while water with low calcium and magnesium content is soft. High levels of hardness produce a build-up of mineral deposits in pipe networks; these deposits are known as scale. Considerably more detergent is required for washing clothes in hard water, and the dishwasher may leave

\\Note:
Surfacings such as asphalt are impervious to water and effectively seal the ground, thus preventing groundwater from being replenished.

\\Note:
The European Union Council Directive 98/83 (EU Drinking Water Directive) concerns the quality of drinking water for human consumption and obliges all member states to implement it stepwise into their national legislation.

Tab.1:
Water hardness ranges

Hardness range	Hardness in mmol/l	Description
1	<1.3	soft
2	1.3-2.5	medium hard
3	2.5-3.8	hard
4	>3.8	very hard

a thin film of lime on the dishes. Water hardness is measured in mmol/l (millimoles per liter). The hardness level depends on the source of the water. › Tab. 1 Water with less than 30 mg/l calcium bicarbonate, on the other hand, does not allow the pipes to form a protective surface layer, with the result that the pipe material is attacked by acids, and corrodes. The effect of water hardness on health is insignificant.

pH

An important measure of the "aggressivity" of water is its pH (Latin: *potentia Hydrogenii*). The pH describes the concentration of hydrogen ions in water, or more precisely: the negative logarithm of the hydrogen ion concentration. On this scale, pure water has a pH of 7, i.e. there are 10^{-7} g H ions in one liter of pure water. If the pH drops below 7, the water behaves aggressively like an acid; if the pH is higher, the water behaves as a base (alkali) and more lime is deposited.

THE DEMAND FOR DRINKING WATER

In the 19th century, Germany required about 30 l of drinking water per day per head for consumption and personal hygiene. Today, by contrast, the figure will soon reach 130 l, due to the increasing levels of sanitary convenience, such as flowing water, showers and flushing toilets. This consumption is doubtless very high, but it has already decreased, because a great number of water-saving fittings have been installed in bathrooms and WCs in recent years. However, industry, commerce and agriculture are using increasing amounts of water. The irrigation of agricultural land consumes the largest quantity of drinking water worldwide.

In the industrialized countries, almost all buildings are connected to the public drinking water supply network. Many billion cubic meters of water are removed from the natural water cycle for drinking water supplies every year. Most of this comes from groundwater and bodies of open water, and the rest from sources such as river bank filtration. The term "bodies of open water" refers to rivers or lakes, the water of which is usually contaminated with bacteria and mechanically eroded solids, and can be supplied as drinking water only after a long purification process.

Conurbations and regions where water is scarce have to rely on some of their drinking water being transported from far away. At the same time, the high proportion of impervious surfaces in cities means most of the rainwater flows directly into their drainage systems. As it is particularly difficult to supply the quantities of drinking water required in these areas, it is imperative to reduce drinking water demand.

The daily drinking water demand of domestic households can be divided into different uses. The amount actually consumed is quite a small proportion of the total. Only about 5 l water are drunk or used for cooking, and the rest used for other purposes. Peaks and troughs during the day are compensated for by water storage at waterworks.

The average hot water demand in domestic residential properties is between 30 and 60 l per person per day. It can vary greatly from day to day and with the habits of the users. A bath requires about 120 to 180 l hot water at 40 °C; a 5-minute shower about 40 l at 37 °C. Energy and drinking water can be saved by choosing to have a shower instead of a bath.

SAVING DRINKING WATER

Today there are many sanitary engineering solutions for saving drinking water: flow limiters in shower heads, water-saving faucets and toilets, and domestic appliances (e.g. washing machines and dishwashers)

Tab. 2:
Typical usage of drinking water

Activity	Usage in l/day/person
Drinking and cooking	5
Basic personal hygiene	10
Baths and showers	38
Dishwashing	8
Cleaning	8
Clothes washing	15
Toilet flushing	40
Garden watering	6
Total	**130**

with reduced water consumption. Installing a water meter in each apartment instead of having one central metering point in the basement has a proven water-saving effect, because users can track their consumption directly; they just pay for the water they have used. WC cisterns with stop buttons and a water usage of 4–6 l per flush are now standard. More advanced systems such as vacuum toilets use 1.2 l water per flush. Composting toilets of various types use no water at all. › Chapter Drinking water systems in buildings, Sanitary rooms

A more accurate analysis of drinking water usage makes it clear that water of drinking water quality is required for only the smallest proportion of the total amount supplied. › Tab. 2 Pure drinking water is necessary only for personal hygiene, washing kitchenware, cooking and drinking. Rainwater-quality water is adequate for toilet flushing, cleaning, or watering the garden. Water consumption can therefore be substantially reduced by using rainwater. Cleaned gray water from showers and hand basins, for example, can also be used for flushing WCs. › Chapter Waste water, Uses of waste water

Merely installing modern water-saving faucets in sanitary rooms can reduce average drinking water demand to about 100 l per person per day. With a few more of the measures mentioned above, it would even be possible to manage on half of normal drinking water consumption with no significant loss of comfort.

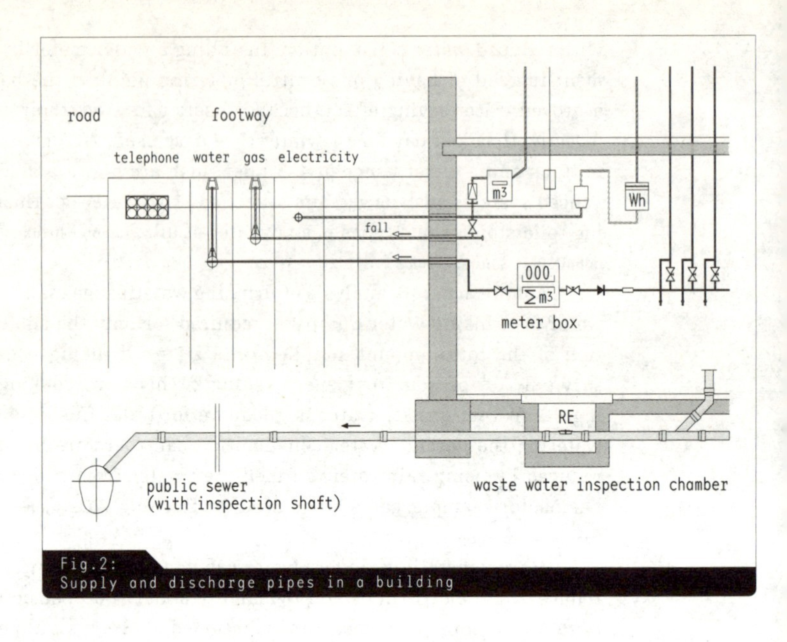

Fig. 2:
Supply and discharge pipes in a building

DRINKING WATER SYSTEMS IN BUILDINGS

The water cycle normally begins in buildings with the supply of cold drinking water through a pipe connected to the public water supply network, unless the plot has its own private supply (well). In larger towns and settlements, the connection to the public drinking water supply is normally at a frost-free depth of between 1.00 and 1.80 m below the sidewalk. Each plot has its own drinking water service pipe, which heads off into the building at right angles to the public supply pipe, as far as the house connection or main stopcock and water metering point. › Fig. 2 In residential properties this pipe has a nominal diameter of about 25 mm (DN 25). › 0

In some European countries, the position of the drinking water connection is marked with a colored sign on a nearby house wall for ease of identification and location of the connection point. The lines and numbers on the sign give the distance to the drinking water connection—from the sign—and the direction (to the right, left, in front or behind). The other abbreviations normally describe the type of connection; the accompanying numbers give the nominal internal pipe diameter.

Water temperature

To prevent microbes from flourishing, the drinking water supplied to the building is cold, i.e. between 5 and 15 °C. To obtain hot water, the drinking water must first be heated in the building. Hot water in this context is described as drinking water with a temperature of between 40 and 90 °C. For personal hygiene, a water temperature of 40 to 45 °C is adequate, while washing dishes requires water at 55 to 85 °C to produce hygienic results.

Water pressure

The water supply company creates a pressure within the water supply network to distribute the drinking water. The water pressure of the public supply is usually between 6 and 10 bar, and is brought down to about 5 bar or less by pressure reducers in the building's pipework system.
› **Chapter Drinking water systems in buildings, Components of a drinking water supply system**

These values may vary from place to place and should only be taken as a guide. The absolute minimum pressure at the draw-off point should not fall below 0.5 bar, as otherwise the water cannot be distributed properly. Pressure losses in the pipework system may be caused by, for example, a great difference in height between the service pipe and the draw-off point. Loss of pressure can be taken roughly as 1 bar for every 10 m height.

COMPONENTS OF A DRINKING WATER SUPPLY SYSTEM

Drinking water is distributed within a building's supply system through a branched network of horizontal and vertical pipes, which are normally concealed in service ducts, wall lining cavities, floor voids, or wall chases. Other water supply system components include a water meter to record consumption, safety devices, stop valves and draw-off points.

House connection

The service pipe to the public water supply and the water meter are part of the supplied service and usually belong to the water supply company. The service pipe takes the shortest route into the building and must not be built over, so that it can easily be located and repaired if necessary. For safety reasons, the pipe duct must pass through the outside wall or foundation of the building at right angles, and the pipe must be enclosed by a protective sleeve.

\\Note:
The abbreviation DN, which is used as a label in every pipework drawing, means "diamètre nominal" and defines the nominal internal diameter of a pipe. This must comply with national regulations and depends on the size of the system.

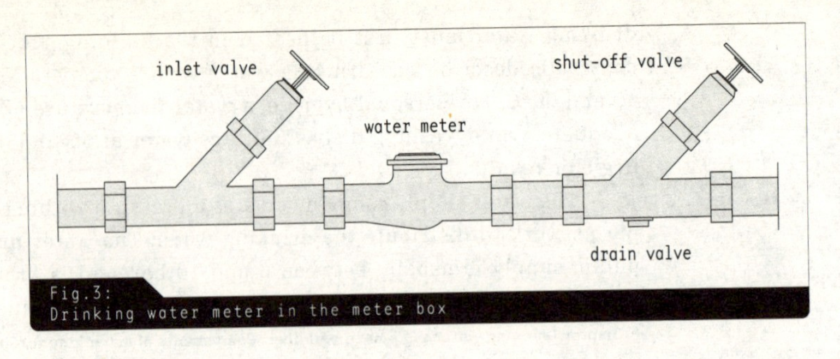

Fig.3:
Drinking water meter in the meter box

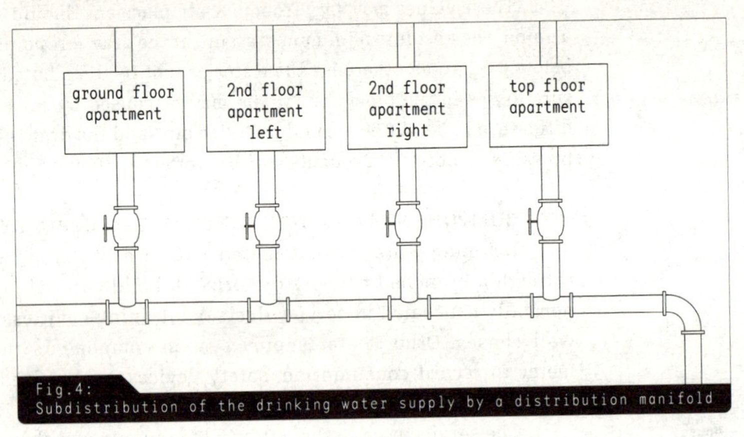

Fig.4:
Subdistribution of the drinking water supply by a distribution manifold

Water meter

The calibrated water meter is positioned directly between the public water supply stopcock and the stopcock for the building's internal water pipework system. These two shut-off valves allow the meter to be removed without complications. › Fig. 3

The architect's design role starts at the water meter. The meter should be housed in a frost-free and readily accessible enclosure, e.g. in a meter box on the road side of the building, so that it can be read easily. If the meter cannot be installed inside the building, it may be housed in a meter chamber outside the building. This may even be specified so as to allow the meter to be read by the water supply company without the building occupant's being present.

Distribution manifold

If the drinking water distribution system has several risers (which may be required, for example, for supplying water to the different floors of an apartment block), a manifold is also incorporated to feed the individual apartments. › Fig. 4 In addition to the subdivision of the drinking water feed, there are often also heating supply pipes, separate draw-off points (e.g. outdoors) and, if necessary, separate pipes supplying water

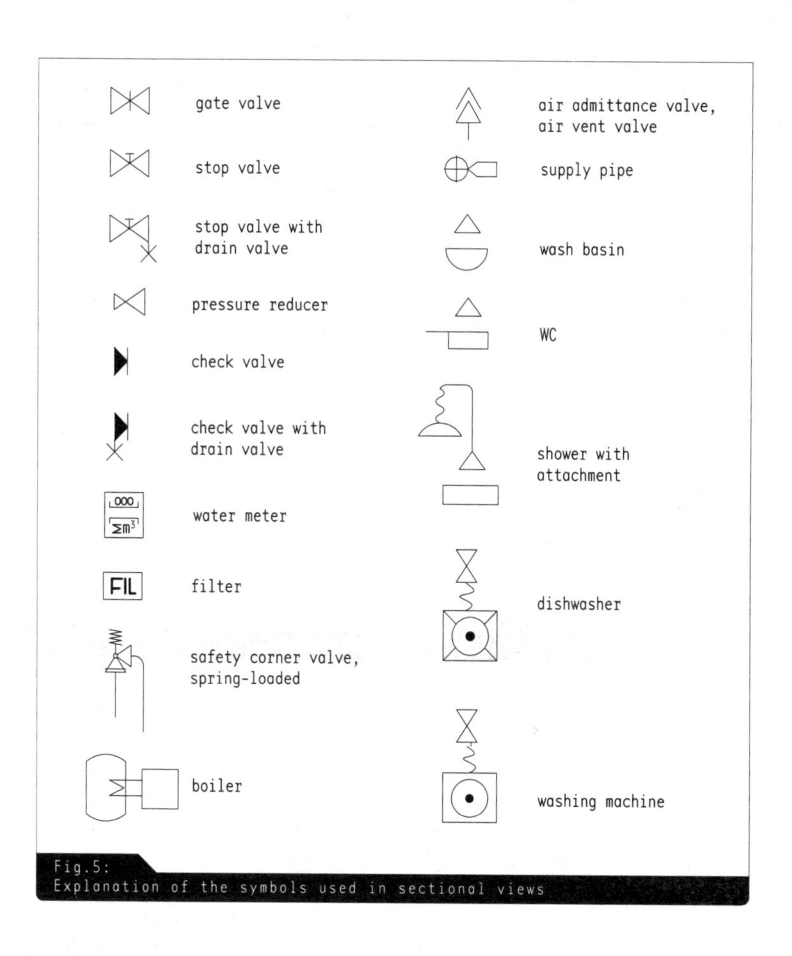

Fig.5:
Explanation of the symbols used in sectional views

for firefighting. Each riser from the manifold is carefully labeled so that it is clear which pipe feeds which premises. Each riser network has its own shut-off valve so that replacement of system components can be carried out independently of the rest of the system.

Layout drawings and symbols

European Standard EN 806 and its national annexes give special graphic symbols for system components and sanitary fixtures for use on design drawings of drinking water systems. They indicate the components to be installed in buildings and their spatial positioning and arrangement. › Fig. 5 These standards may vary from country to country. The system must be drawn in plan and sectional views to fully depict the drinking water system and its associated pipework. Since sanitary appliances in plan are

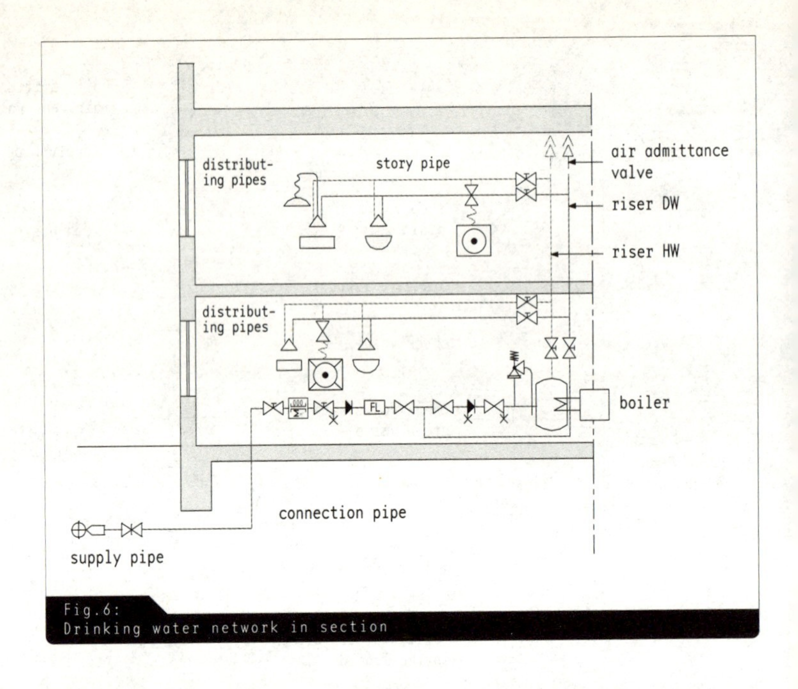

Fig.6:
Drinking water network in section

seen from above, some appliances may require different symbols for plan and sectional views. A key on the drawing is generally helpful to explain the symbols used.

Representation in section

The representation of the system in section should be schematic and contain as much information about the drinking water system as possible. The sequence of the drinking water system symbols used should correspond to the actual arrangement of the sanitary items. Symbols and pipework are drawn as if the supply network and draw-off points were in the same, single plane. › Fig. 6

Representation in plan

When drawing the system in plan it is important to identify the rising and falling pipes separately with arrows showing the direction of flow in the pipes. › Fig. 7 You should also show whether a pipe begins, ends or carries through each story. › Fig. 8

Description of pipework components

Many pipework components in drinking water systems are referred to by special terms that are not used in other supply systems. › Fig. 6 The most important terms are:

_ service pipe for the pipe between the public supply and main stopcock in the building

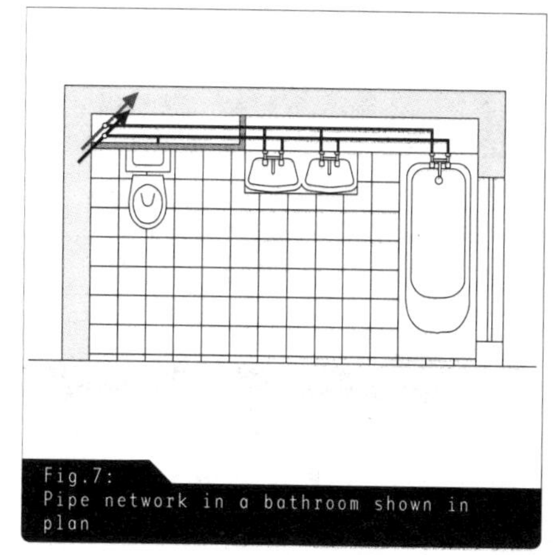

Fig.7:
Pipe network in a bathroom shown in plan

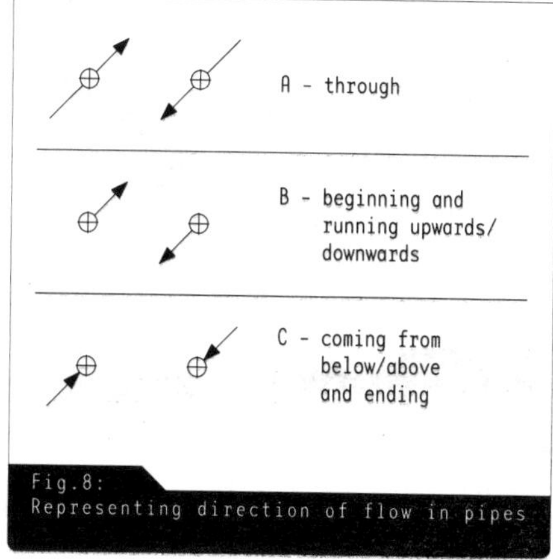

Fig.8:
Representing direction of flow in pipes

- riser for the pipe which passes vertically through the building, from which the
- pipes supplying the individual stories branch off horizontally
- circulation pipe, which provides continuous hot water at the draw-off points, but is not always required
- distributing pipes are the vertical pipes branching from the horizontal pipes supplying each story to the draw-off points.

The internal diameter of risers in residential buildings is about DN 20, and for pipes supply stories about DN 15, i.e. nominal diameters of 20 and 15 mm respectively.

Circulation pipes

A circulation pipe ensures that hot water is available immediately at draw-off points. This has the advantage that a large amount of cold water does not have to flow through the pipe before the hot water appears, which is frequently the problem with instantaneous water heaters. A disadvantage is the electrical energy continuously consumed by the pump to keep the water circulating in the pipework. A time-switched pump, which only runs when hot water is needed, can mitigate the effect.

Pipe routes

Horizontal pipe routes may pass under a basement ceiling or be placed in the floor construction. Larger buildings often have floor ducts, or position the pipes above suspended ceilings. › Fig. 9 Vertical pipes in basements or equipment rooms may be fixed openly on the walls, › Fig. 10 and in the stories above in installation shafts or, for short lengths, be concealed in half-height false walls. › Chapter Drinking water systems in buildings, Sanitary rooms

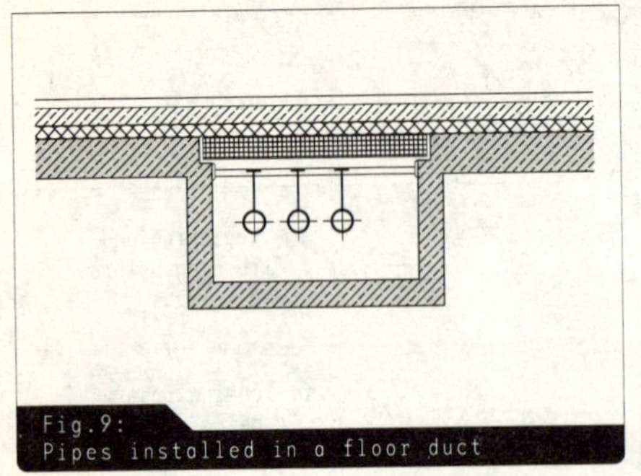

Fig.9:
Pipes installed in a floor duct

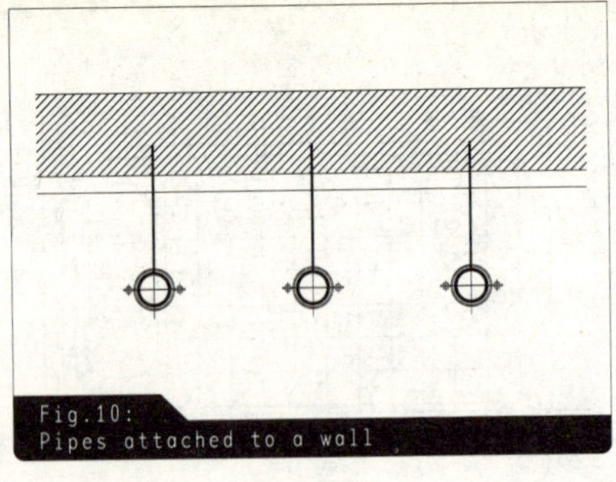

Fig.10:
Pipes attached to a wall

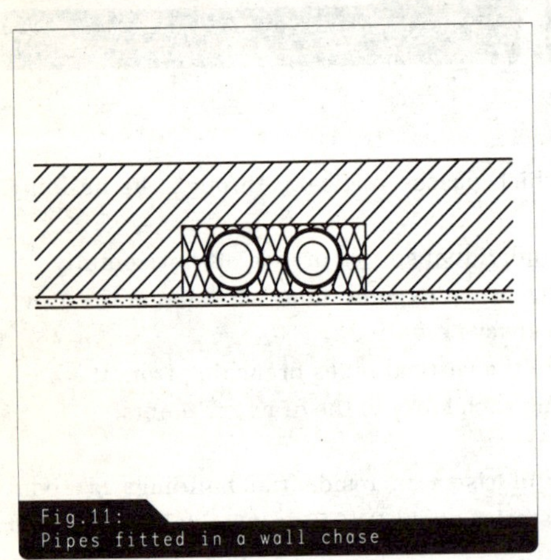

Fig.11:
Pipes fitted in a wall chase

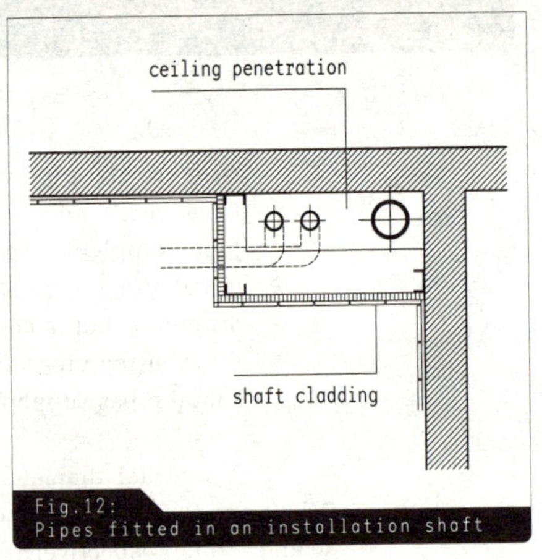

Fig.12:
Pipes fitted in an installation shaft

In solid wall and floor construction, pipes are fitted into insulated wall chases, if the wall has a large enough cross section and there are no structural engineering reasons to preclude it. › Fig. 11 This method of installation is increasingly being superseded by installation shafts in sanitary rooms because of its complexity and poor sound insulation. › Fig. 12 False walls, in comparison with shafts, just clad the pipes in the wall cavity and generally terminate at half room-height, while an installation shaft can carry pipes through several stories. › Fig. 13

Calculation of pipe sizes

The nominal diameter of the pipes is determined by the number of consumer points connected. › Tab. 3 It is also important to consider the probability of simultaneous draw-off, the material used for the pipes, the

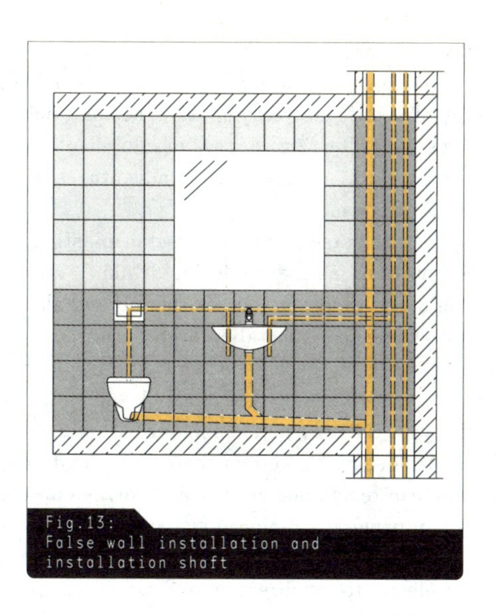

Fig.13:
False wall installation and installation shaft

Tab.3:
Diameter of drinking water pipes

Pipe type	Approximate internal diameter
Service pipe	DN 25 to DN 32
Riser	DN 20
1-5 draw-off points	DN 20
5-10 draw-off points	DN 25
10-20 draw-off points	DN 32
20-40 draw-off points	DN 40
Story distribution pipes	DN 15
1 WC cistern	DN 10 to DN 15
1-2 wash basins	DN 15
1 shower	DN 15
1 bath	DN 20 to 25
1 garden hose	DN 20 to 25

pressure loss due to friction, and minimum flow pressure. The system must be designed to have a minimum flow pressure in the pipe such that the remotest drinking water draw-off point always has sufficient pressure to operate properly. The pipe friction pressure loss factor is the fall in pressure within a section of pipe divided by its length.

Materials

Modern drinking water pipes that are in the ground outside the building are normally of plastic (polyethylene HD), because metal pipes require additional corrosion protection. A new development is the multi-layer metal composite pipe, which combines the advantages of metal (strength) and plastic (corrosion resistance).

Copper, galvanized steel, stainless steel or plastic (polyethylene) are used inside the building as materials for drinking water pipes. Plastic pipes are not rigid and their small cross section and ability to be bent to tight radii allows them to be installed even within a floor construction. Polyethylene pipes are usually designed as a pipe-in-pipe system: the flexible drinking water pipe (PE-X) is surrounded by an additional flexible, external protective pipe (PE-HD), and can be taken out of the protective pipe and replaced if necessary. As well as their high flexibility, plastic pipes have the advantage of resistance to scale and corrosion. When metal pipes are used or when pipework is replaced as part of a refurbishment project, attention should be paid to placing the less noble material after the more noble material (relative to the direction of flow), e.g. copper after steel, as otherwise corrosion could occur.

Safety devices

Most systems incorporate safety measures to maintain the high quality of their drinking water. The quality must not be reduced by the entry of non-potable water, for instance. These devices include safety fittings that prevent back-flow or suction of contaminated water and the mixing of drinking water with water of lesser quality, which may happen, for example, if there is underpressure in the system due to a pipe developing a leak at the same time as a shower head has been left in bathwater. These safety devices prevent the underpressure from sucking bathwater into the drinking water supply.

\\Note:
Polyethylene HD is also used for buried pipes; PE-X is used mainly for internal pipework with a protective outer covering of PE-HD. HD means high density, PE-X crosslinked polyethylene.

\\Note:
The maintenance of purity of drinking water in buildings is governed by European Standard EN 1717, "Protection against pollution of potable water in water installations and general requirements of devices to prevent pollution by backflow".

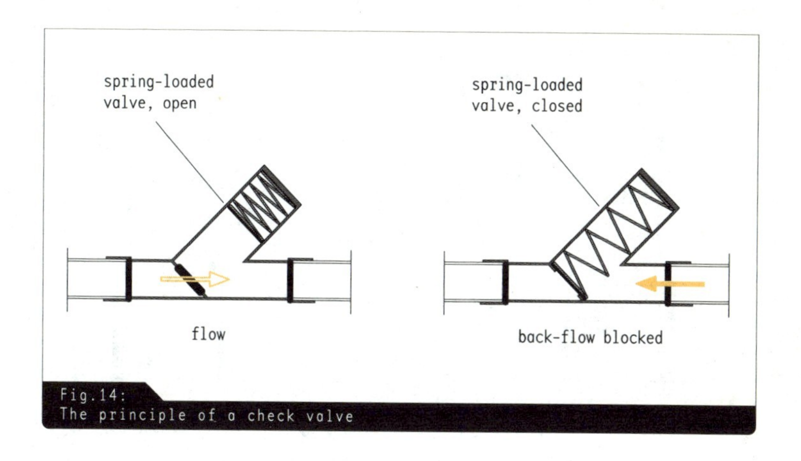

Fig.14:
The principle of a check valve

Check valves

One of the safety devices normally built into a domestic water installation is a check valve placed after the water meter. It is a self-closing, spring-loaded valve that prevents the backflow of drinking water in the pipe. The check valve opens only to allow water through in the correct direction of flow. If the flow stops, it closes again. If the direction of flow reverses, the valves closes with increased pressure. › Fig. 14

Air admittance valve

Check valves are normally installed in combination with an air admittance valve, which acts to compensate any underpressure occurring in the pipework system and prevent back-suction of contaminated water into the drinking water supply. The air admittance valve is positioned at the highest point in every cold or hot water riser. The valve inside the air admittance valve is normally closed. › Fig. 15 In conjunction with the check valve, the air admittance valve opens when underpressure occurs, and the inward flow of air prevents used water from being sucked back into the pipework. › Fig. 16

As there is a risk of water escaping from the air admittance valve when there is overpressure in the pipework, the valve may be connected to a drip collection pipe, which drains any escaping water into the waste water pipework and thus into the sewer. A drip collection pipe is not required if the air admittance valve is positioned above a shower or basin, where any escaping water cannot cause any damage.

Pressure reducer

The water pressure provided by the water supply company may be too high for normal draw-off points and may need to be reduced by a pressure reducer inside the building. The delivery pressure acts on a moving diaphragm, which either opens or closes a connected spring valve, depending on the setting. › Fig. 17 The pressure reducer should be installed in a position where it can be easily maintained.

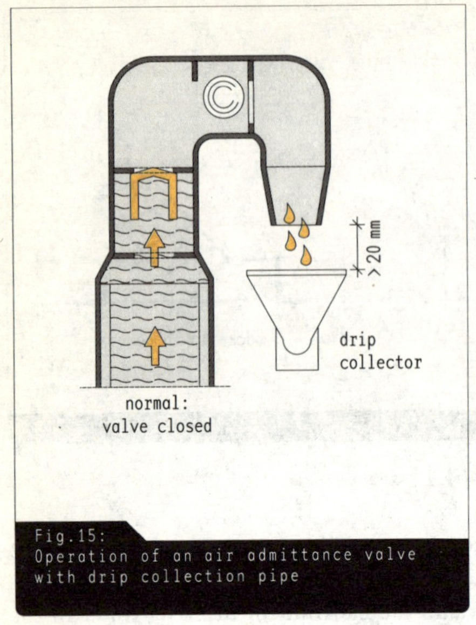

Fig.15:
Operation of an air admittance valve with drip collection pipe

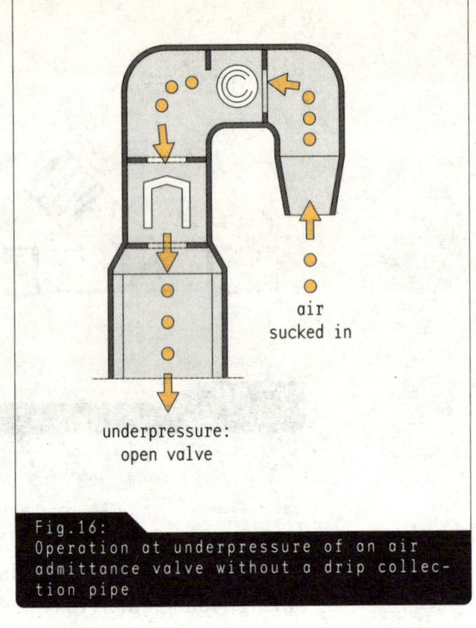

Fig.16:
Operation at underpressure of an air admittance valve without a drip collection pipe

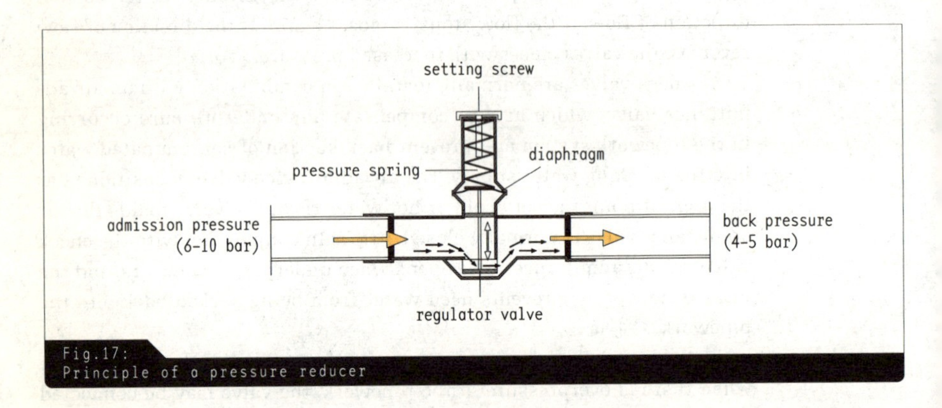

Fig.17:
Principle of a pressure reducer

Filter

A fine filter is built into every drinking water system to ensure that it is free of dirt and rust particles. In most cases it is installed between the water meter and the pressure reducer, so that the latter is not contaminated. Installing a filter is worthwhile only if it is regularly serviced, because a filter insert cannot be expected to remain permanently free of microbes.

Safety distances

Buried drinking water pipes should be placed more than 1 m from any wastewater pipes above them, so as to prevent contamination of the

Tab.4:
Typical insulation thicknesses for cold water pipes

Pipe	Insulation thickness
_ In the open in an unheated space _ In an installation shaft with no nearby pipes carrying hot water _ In a wall chase	4 mm
_ In the open in a heated space	9 mm
_ In an installation shaft with nearby pipes carrying hot water	13 mm
The insulation thickness is calculated assuming a thermal conductivity of 0.035 W/m^{2}K and must be recalculated for other insulation materials	

drinking water supply in the event of a leak. If it is not possible to maintain this safety distance, the drinking water pipe must be laid above the waste water pipe, at a distance of at least 20 cm.

Thermal insulation

Cold drinking water pipes should be prevented from inadvertent heating from nearby hot water or heating pipes by insulating them, or by maintaining a suitable distance to keep them free of microbes. Drinking water pipes should also be insulated when they pass through heated spaces. › Tab. 4

HOT WATER SYSTEMS

To satisfy the daily demand for hot water, part of the drinking water in a building is heated before being distributed. Hot water supply systems consist of the cold water feed, a boiler, perhaps a hot water storage tank, hot water distribution pipes leading to draw-off points, and in some circumstances circulation pipes, which ensure that hot water is instantly available at the draw-off points.

\\Important:
Cold water pipes should always be fixed below gas pipes, as there is always a risk that condensing water could cause corrosion of the gas pipe and result in a dangerous gas leak.

Tab.5:
Conventional insulation thicknesses for hot water pipes

Nominal pipe diameter	**Insulation thickness**
_ up to DN 20	20 mm
_ from DN 22 to DN 35	30 mm
_ from DN 40 to DN 100	pipe DN
_ larger than DN 100	100 mm

Half the above thickness is adequate for pipe lengths of up to 8 m
_ at wall and ceiling penetrations
_ where pipes cross

Pipe layout

If a building obtains its hot water from a central installation, the hot water pipes mainly run parallel to the cold water pipes throughout the building and have more or less the same pipe cross section. The water temperature in the pipework is between 40 and 60 °C. To avoid energy losses, the hot water pipes should be kept as short as possible and always thermally insulated where they pass through unheated spaces. The extra space required for insulation must be taken into account when designing the pipework route.

The insulation thickness is approximately equal to the pipe diameter; with pipes less than 8 m long, half that thickness is adequate. This is also the case where pipes pass through walls or ceilings and where they cross, e.g. in the floor construction. › Tab. 5

Individual, group or central hot water supply

Hot water systems can be central or decentralized in the building or provide hot water directly at the draw-off point. If a hot water source has only one draw-off point connected to it, this arrangement is called an individual supply, while with several connected draw-off points it is a group supply. › Figs. 18 and 19 With a central supply all draw-off points are supplied with hot water from a single, central boiler. › Fig. 20 It is also possible to combine individual, group and central systems, for example to switch off a central boiler in summer and yet have hot water available through individual supply points.

Boilers

There are basically two types of boilers used to provide hot water: continuous flow or instantaneous water heaters, which heat the water directly as it is used; and storage water heaters, which keep the water constantly hot and ready for drawing off. A further difference between the types relates to the source of heat. Heat sources include solid fuel, oil, gas, electricity, geothermal, and solar energy. Heating takes place either indirectly through a heat exchanger and a heat medium, or directly by the application of heat to the water to be heated.

› 🖇

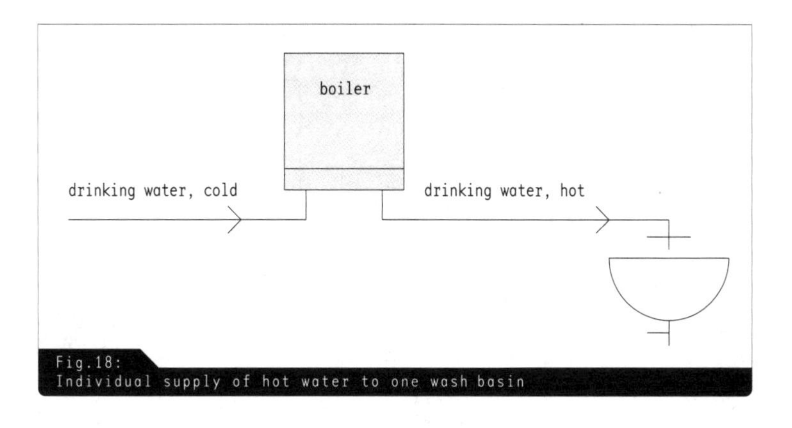

Fig.18:
Individual supply of hot water to one wash basin

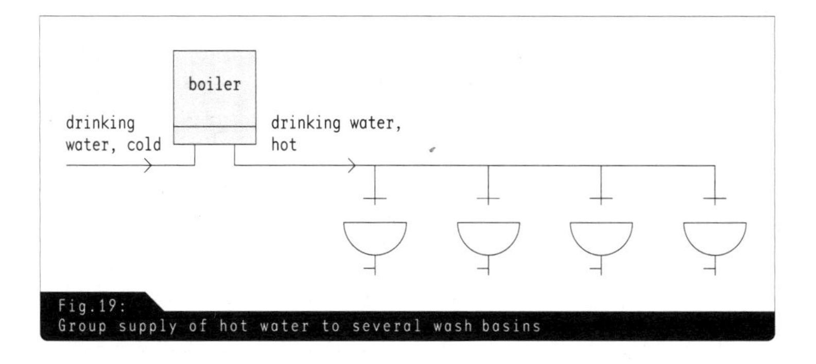

Fig.19:
Group supply of hot water to several wash basins

If possible, boilers installed in a central plant are used to heat the building and supply it with hot water. The plant consists of a hot water storage vessel, which stores service water and provides a source of heat for the heating circuit, and a connected boiler, which releases its heat to the service water. Hot water and heating energy are transported

\\Note:
A heat exchanger is used to transfer heat from one medium into another. The heat medium could be water, which releases its heat into the air, for example, as with a radiator. In hot water supply systems the heat exchanger is inside the hot water storage vessel.

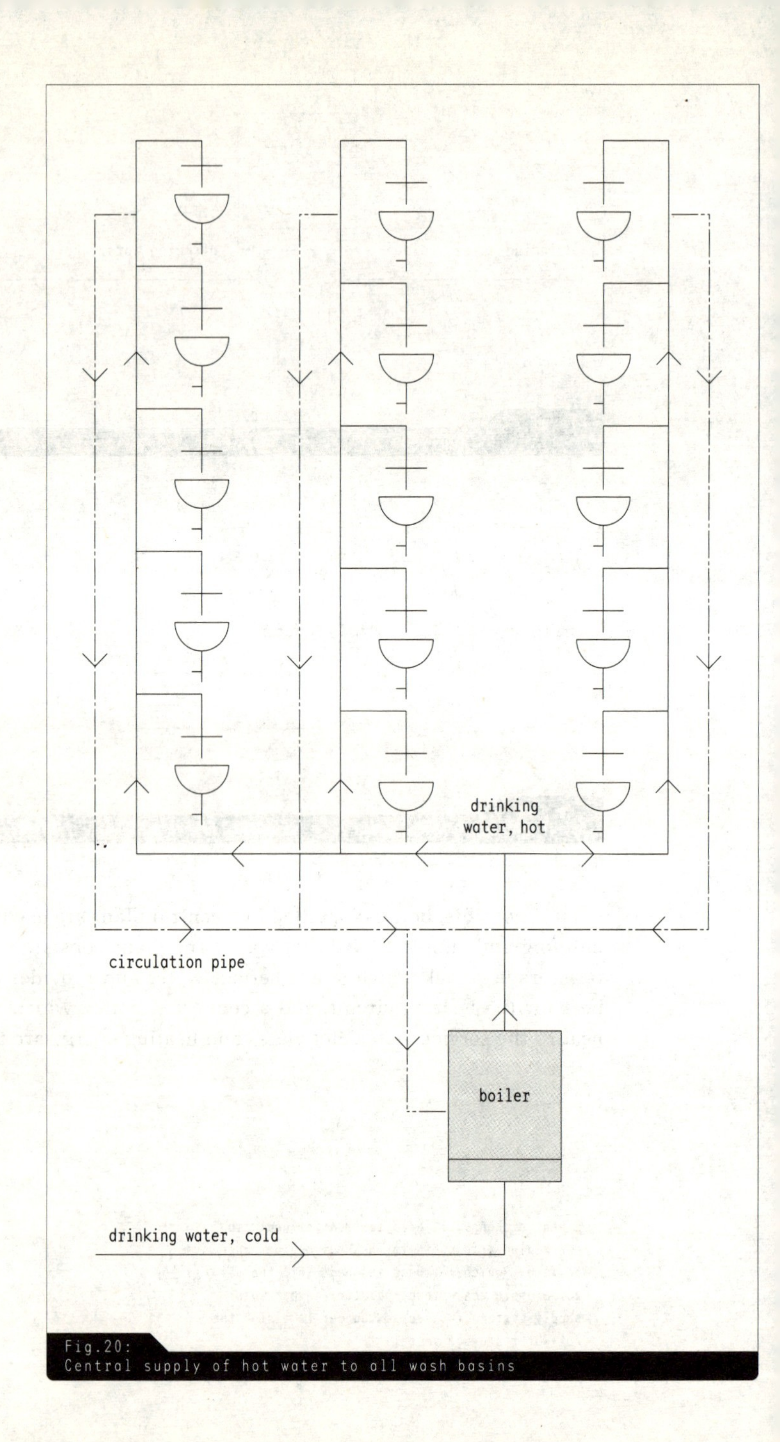

Fig.20:
Central supply of hot water to all wash basins

by pumps through the pipe network to the draw-off points and heating radiators. Central plants have the advantage that a solar thermal energy installation can be connected or retrofitted at a later date and contribute to supplying hot water to the hot water storage vessel.

Continuous flow systems

Continuous flow systems, also called instantaneous or continuous flow water heaters, heat the water directly to a temperature of about 60 °C. › Fig. 21 Their advantage is that only the amount of hot water needed is heated. In contrast to storage systems, there are no standby heat losses with continuous flow systems, and the water can generally be classed as fresh.

Continuous flow water heaters are inexpensive to install and save space. They are highly efficient because they heat the water directly. The startup phase cannot be avoided: hot water can be supplied to the draw-off point only after a delay, which results in fresh water flowing unused with the waste water into the sewers. The pipes from the instantaneous water heater to the draw-off points should be as short as possible, so that this cold water phase is curtailed. It cannot be completely eliminated, however.

Instantaneous water heaters can be operated with electricity or gas. Electrical systems generally use an AC supply, while the gas systems require a flue or chimney connection. Electrically powered instantaneous water heaters are associated with high energy costs. As electricity is very expensive to produce, its use as an energy carrier should be restricted to providing small quantities of water or to situations when no other energy carrier can reasonably be considered. Gas, on the other hand, although it is a fossil fuel, produces the least amount of carbon dioxide (CO_2) of all the fossil energy carriers.

If designed as gas combination boilers, instantaneous water heaters can supply hot water and heating energy at the same time in a suitable system. Water quantities can be controlled hydraulically, thermally or electronically. Continuous flow systems can serve several draw-off points

\\Note:
Coal, oil and gas are fossil fuels. When burned they produce carbon dioxide (CO_2) and contribute to global warming. There will also come a time when they are finally used up and are no longer available as energy sources.

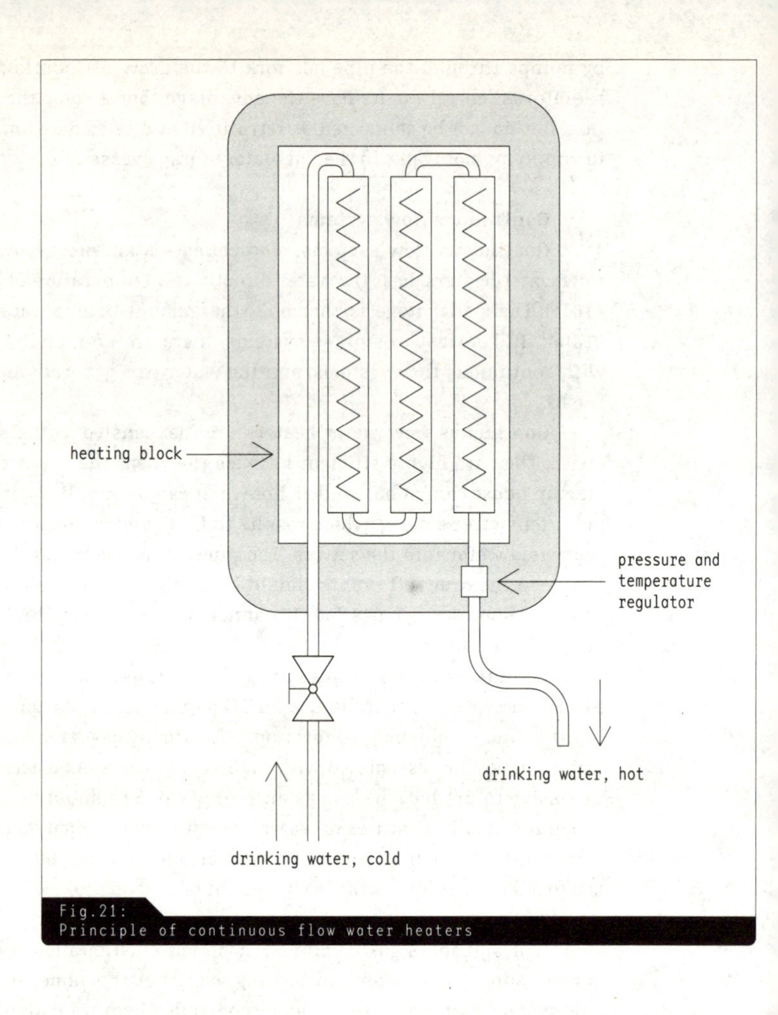

Fig.21:
Principle of continuous flow water heaters

simultaneously and are used for individual, group and central hot water supplies. Maximum water flow is limited, however; when there is a high simultaneous draw-off, such as might be the case for hotels or sports halls, the output of a continuous system is too low. In this case a central storage system should be used.

Instantaneous water heater with integrated water storage

Another hybrid form is an instantaneous water heater with a small, integral hot water reservoir holding between 15 to 100 l water. If more water is needed, e.g. for a bath, the rest of the bathwater is heated as if by an instantaneous water heater. Continuous flow systems with integrated

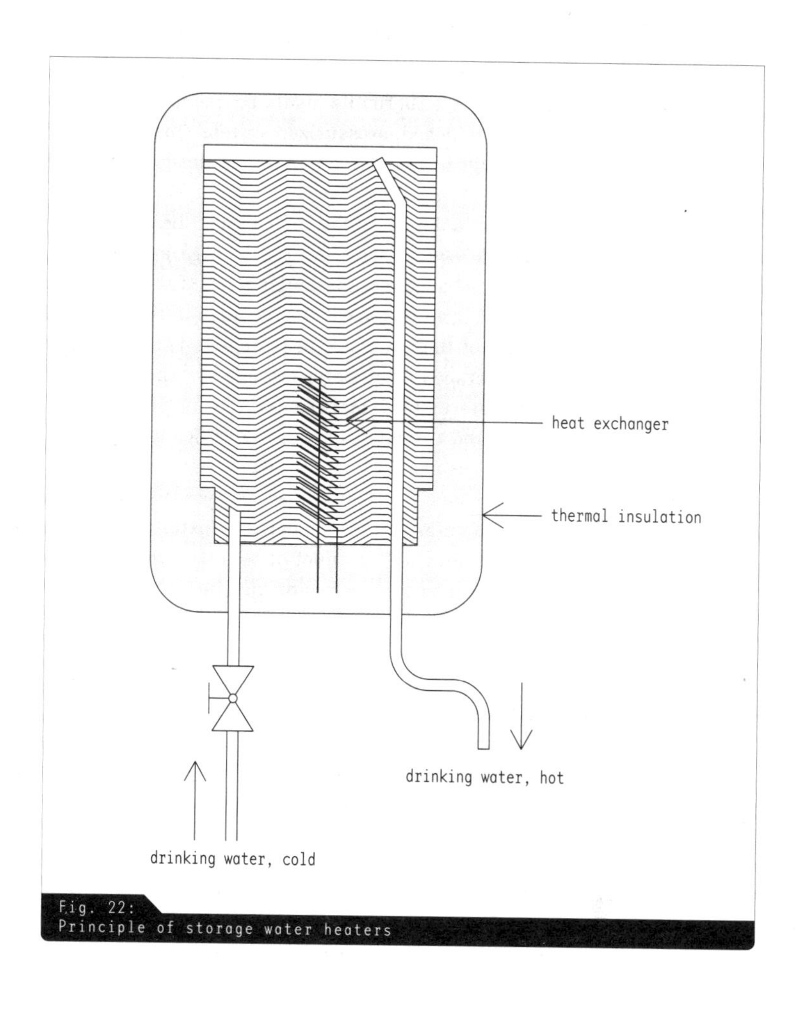

Fig. 22:
Principle of storage water heaters

storage are mainly used to supply smaller housing units with hot water and heating energy.

Storage systems

These storage water heaters continuously heat the hot water stored to keep it at a constant temperature of about 60 °C. › Fig. 22 The water is heated directly by a connected heat source or indirectly by a heat carrier, which could be an antifreeze solution in a thermal solar installation. › Chapter Hot water systems, Solar heating of water

Hot water storage must be located centrally in the building near the heating plant. It is thermally insulated and is able to supply several draw-off points as a closed, pressurized system. Open, unpressurized and uninsulated storage systems like boilers are designed for individual draw-off points.

Thermal solar installations can also be connected or retrofitted to hot water storage systems. As the storage and boiler are directly connected in these systems, the boiler reheats the water if the temperature rise achieved from the thermal solar installation is insufficient, or if the temperature of the hot water in storage falls below a particular set level. The disadvantage of hot water storage systems is that the water may go stale if it is stored for long periods.

Compared to instantaneous water heaters, which are located close to the points of use, central storage systems are associated with higher installation costs, as their pipework systems are usually considerably longer and therefore more expensive to install. However, overall costs can be reduced if hot water production is linked with heating the building, as it means only one boiler has to be installed.

Legionella

Oversized pipework systems, drinking water heating systems with large storage capacities or poorly insulated drinking water pipes provide the right conditions for *Legionella* to multiply. *Legionella* are rod-shaped bacteria that are present in cold water in low concentrations but multiply quickly in warm water. Humans become infected not by drinking the water but by inhaling the aerosol created by agitating the water, for example, when showering. The symptoms of the illness are like those of a lung infection, starting with a fever, muscle pain, cough, and severe shortness of breath. It is very easily mistaken for influenza-type illnesses. If the illness is not diagnosed in time and treated with a suitable antibiotic, it can be fatal.

If the water temperature in a hot water storage vessel falls to between 30 to 45 °C for an extended period, there is the risk of *Legionella* contamination. A simple but effective method of preventing the build-

\\Note:
Unpressurized, uninsulated storage systems could be boilers or point-of-use water heaters, which are mostly installed as undersurface appliances below basins or sinks. They are suitable for producing small quantities of hot water quickly, e.g. in office kitchenettes.

up of *Legionella* in drinking water is to thermally disinfect the stored water. This can be done by raising the temperature of the hot water to over 60 °C, which kills the bacteria, daily or weekly. Another method is electrolytic disinfection, which works by creating disinfecting agents in the water.

Solar heating of hot water

Solar collectors provide the most environmentally compatible form of heat energy because it is not accompanied by any emissions. Thermal solar installations are primarily used for the provision of hot water. Depending on the climate and given a favorable building, if the area is doubled the installation can also be used to support the heating system.

Thermal solar installations consist of flat plate or vacuum tube collectors, which differ in their efficiency and manufacturing cost, the fluid circuit for transporting the generated heat using a water-glycol mixture, and the hot water storage vessel in which the water is heated. A heat exchanger in the hot water storage vessel transfers the heat transported from the collector into the water in the vessel. › **Fig. 23**

The design of the solar installation depends on the selected collector type and whether the intended use is hot water provision alone or also to support space heating.

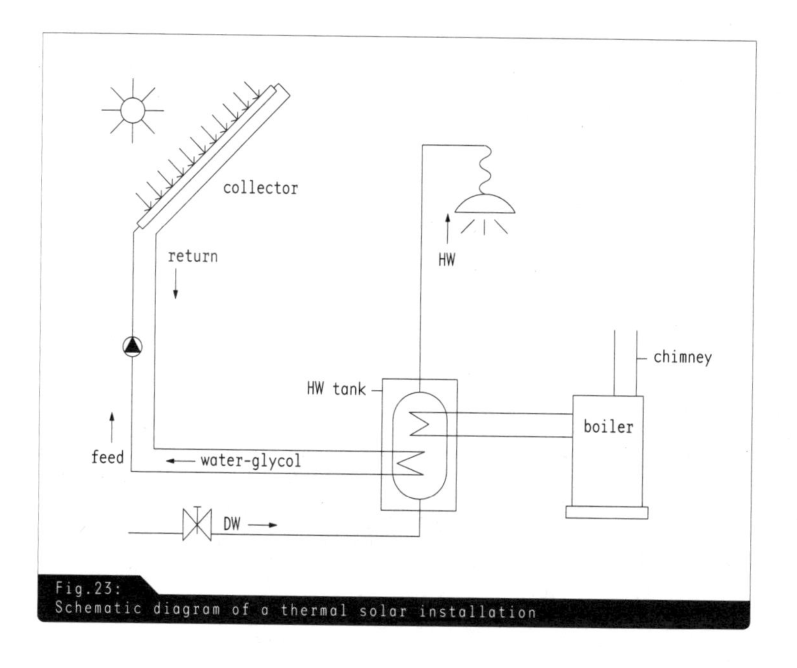

Fig. 23:
Schematic diagram of a thermal solar installation

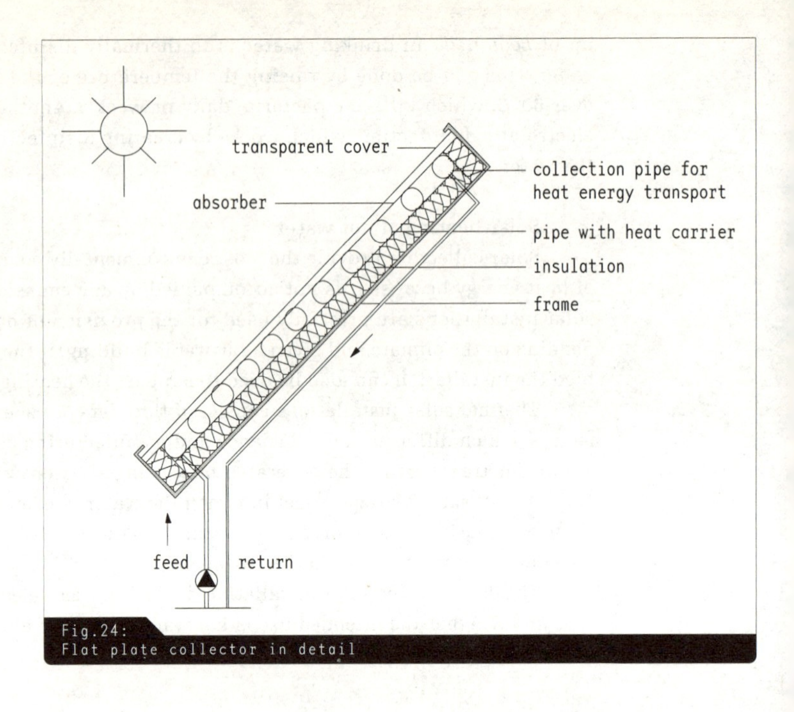

Fig.24:
Flat plate collector in detail

Flat plate collectors

Flat plate collectors consist in principle of an absorber layer with a highly selective coating, which allows the collector to absorb almost all of the solar radiation that falls upon it. The absorber is covered with a transparent cover of non-reflective safety glass with a high solar energy transmittance value and thermal insulation on the back and sides. The whole system is supported on a frame. Pipes filled with a heat transfer medium under the glass cover absorb the heat and transport it to the hot water storage system. › Fig. 24–26

\\Tip:
In a central European climate, a flat plate collector surface area of between 1.2 and 1.5 m^2 per person would be enough to provide hot water to residential buildings. Double that area would be needed to provide support to the space heating system in winter as well: 2.4 to 3.0 m^2 per person.

Fig.25:
Flat plate collector installed on a roof

Fig.26:
Flat plate collector integrated into a facade

Fig.27:
Vacuum tube collector

Vacuum tube collectors

With vacuum tube collectors, the absorber layer is inside airtight glass tubes to increase efficiency. › Fig. 27 The collectors consist of several glass tubes arranged close to one another and connected by a special mount to the collection tube, which is filled with a water-glycol mixture. Vacuum

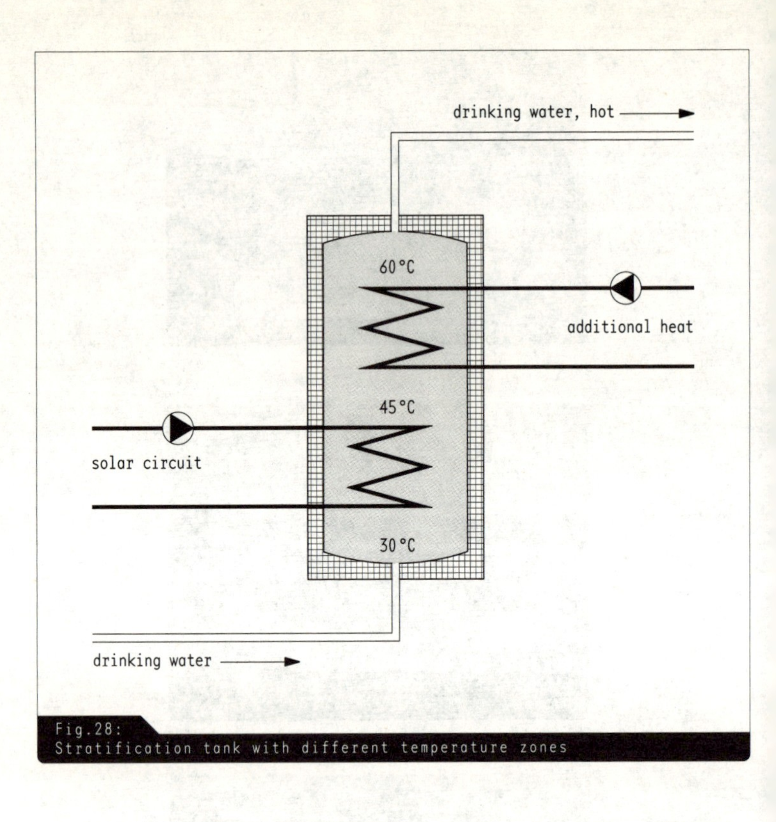

Fig.28:
Stratification tank with different temperature zones

tube collectors can be set at a particular angle by rotating them at the time of installation, so that a good yield can be obtained, even if the unit is at a relatively unfavorable vertical angle or less than optimum alignment. Metal plate reflectors attached to the sides can increase the amount of solar radiation the unit receives. The higher efficiency of vacuum tube collector systems means that an area of 0.8 to 1.0 m² per person is required for hot water provision in residential buildings.

Alignment

Solar collectors are usually integrated into a southeast or southwest-facing pitched roof, or set at a vertical angle of 30 to 45° on a flat roof if they are intended to provide hot water all year round. If they are to provide support for space heating in Central Europe, the angle should be up to 60° as the winter sun is considerably lower in the sky. The units, in particular those with vacuum tube collectors, can also be attached in front of balconies, facades or similar alternative positions. If, for constructional reasons, the units cannot be installed at a suitable vertical angle or if the roof is not quite facing in the right direction, then a greater collector area should be provided.

Degree of coverage

If the amount of available solar radiation in winter is small, collectors can cover only part of the annual hot water demand and an additional means of water heating is required. One possibility that could be recommended is a combination of hot water provision and support to space heating. This system stores water in zones of different temperature in a stratification tank and can be used as a heat source for space heating. › Fig. 28

In a stratification tank, the hottest water is at the top over a middle mixed zone, with the coolest water at the bottom where cold drinking water is introduced. The feed to the heating circuit is taken from the top of the stratification tank, where higher temperatures prevail. When there is insufficient solar radiation, a separate boiler provides the additional heat through a heat exchanger. In this way about 25% of the heating energy demand can be fulfilled from renewable sources. Solar installations supporting space heating are most effective using a combination of floor or wall heating because this method of heating requires lower feed temperatures than heating radiators.

SANITARY ROOMS

Hot water is produced by one of the systems described above, and distributed in parallel with cold water in a system of pipes to be ready for use at various draw-off points in kitchens, sanitary and other rooms with a hot water requirement. Sanitary rooms are mainly used for personal care and hygiene. Of all the rooms in a house, they are the most intensive in their use of building services installations, as they have cold and hot water supply and the associated waste water disposal systems. Depending on their design, they may be termed rooms with wet areas or wet rooms. To cut down the work involved in installing the pipework, the layout of these rooms should be chosen to reduce the number of installation shafts and shorten the water supply and disposal pipes, as far as possible. Grouping the pipes together not only simplifies the plumbing work, but also reduces the transmission of sound to neighboring rooms.

\\Note:
Further information on the heating of buildings can be found in the chapter on "Tempering systems" in: Oliver Klein and Jörg Schlenger, *Basics Room Conditioning*, Birkhäuser Verlag, Basel 2008.

Sound insulation

Noise from sanitary rooms often originates from WC cisterns, water flowing in waste pipes, faucets drawing off water, or activities in the room. These sounds are transmitted to other rooms through walls, ceilings and floors. Quiet rooms or bedrooms should not be positioned adjacent to bathrooms or toilets if disturbing noises are to be avoided. Walls with services in them next to bedrooms, for example, cannot be insulated to the extent that they emit no sound at all. Siting sanitary rooms next to separating walls between residential properties is only recommended if there is also a kitchen, bathroom or other room where noise is not an issue on the other side of the wall in the adjoining property, unless there is an acoustic isolation joint between the two parts of the building.

Wall chases are not advantageous in terms of sound transmission. Surrounding the pipes in the chases with insulation will certainly attenuate the noise, but the chase then has to be cut deeper, which is usually associated with structural stability problems.

Baths and lavatory pans standing on the floor should be bedded on an elastic isolating layer or a floating screed, so that the noises from them are not transmitted through the floor into neighboring rooms. Wall-hung sanitary appliances such as lavatory pans, wash basins or shelves should be attached to walls with a high mass per unit area using sound-insulating sleeves or plastic profiles or attached to false walls.

Faucets and valves in bathrooms are manufactured in two noise categories: low-noise faucets are category I; those that emit higher levels of noise are category II. For noise insulation, category I faucets are preferable, although they may sometimes give rise to higher costs.

The design of sanitary rooms is always a difficult task for architects, as it involves not only the layout and style but also the sound insulation and the integration of extensive pipework. This calls for great attention to detail, as a poorly thought-out arrangement of sanitary appliances and the resulting awkward pipework routes often cause technical, functional, and financial problems.

Arrangement of sanitary installations

When planning a sanitary room, an architect must take into account how far the sanitary appliances are from the drinking water risers and waste water stacks, and how directly and simply the connections can be made. While drinking water pipes generally have a small cross section and can even be installed in the floor structure without much problem, waste water pipes are more difficult to incorporate because of their relatively large diameters and required fall of 2% within buildings. Depending on the type of sanitary appliance, waste water pipes often start off slightly higher

Fig. 29:
Substructure of a double-sided false wall

than floor level, which means connection to the stack is easy provided the distance is short.

Exposed drinking and waste water pipes create high levels of noise. On the other hand, false wall installations of various constructional types, or shafts that pass from story to story and conceal pipework, increase sound insulation and dispense with the need for expensive wall chases.

False wall installations

Instead of attaching sanitary appliances to a solid wall and forming the void around the pipe route with conventional masonry, most pipes are now installed behind false walls to preserve the structural stability of the main walls and provide better sound insulation. They consist of a metal supporting frame and a system for attaching the sanitary appliances. The remaining space inside them is filled with insulation and the frame is clad with plasterboard. › Fig. 29 False walls are normally between 1.00 m and 1.50 m high and are fixed about 20 to 25 cm in front of the real wall, depending on the diameter of the pipes behind them. They conceal only the pipework for that particular story and not pipes from other stories unless they are directly connected to an installation shaft. The top surface of a false wall can be used as a bathroom shelf.

Another version of a false wall installation is the modular assembly block system. This involves prefabricated, compact elements formed from polyester foam concrete, which encapsulate all the supply and disposal pipe connections, inbuilt flushing cisterns and all the fastenings for the sanitary appliances to be connected to them. They are about 15 cm deep and are either impact-sound insulated at their connections to the wall, or stand on supports on the structural floor. The cavities must be walled up or filled with mortar.

Fitting out sanitary rooms

The size and fitting out of sanitary rooms depends primarily on the number of occupants and their particular needs. The room dimensions of a bathroom, on the other hand, depend mostly on the sanitary appliances to be installed and the required distances between them. › Fig. 30

A separate bathroom and toilet arrangement should be considered for residential units with more than two occupants to allow more convenient everyday use. For a family with more than two children, there should be another shower in addition to the actual bath, as well as a separate WC. In all considerations of the size and position within the building of sanitary rooms, the designer should take into account that rooms with wet areas should generally be close to one another, so that pipes can be installed in groups and long pipes are not necessary to transport the water.

Sanitary appliances need to be spaced a minimum of 25 cm from one another or placed at different heights to allow them to be used without interference. Hence a washbasin, for example, may be positioned with an edge extending laterally over a bath at a lower level. Similarly, there should always be a clear area in front of each sanitary element to ensure freedom of movement. › Fig. 31

Wash basins

Wash hand basins and wash basins differ in their sizes. While small wash hand basins are found in lavatories and are intended for hand washing only, wash basins have larger dimensions. They should allow an arm to be immersed up to at least the elbow. Most are manufactured from sanitary ceramics or acrylic; in a minority of cases from enameled or stainless steel. The top edge of these appliances is normally between 85 and 90 cm above floor level. Double wash basin units are more economic in their use of

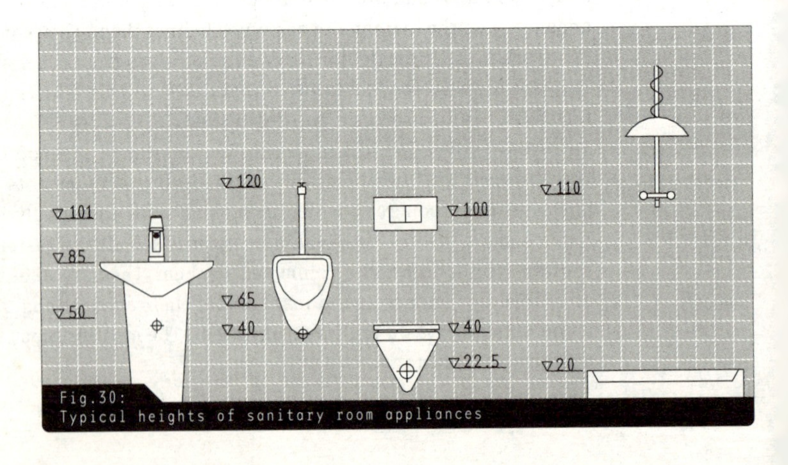

Fig.30:
Typical heights of sanitary room appliances

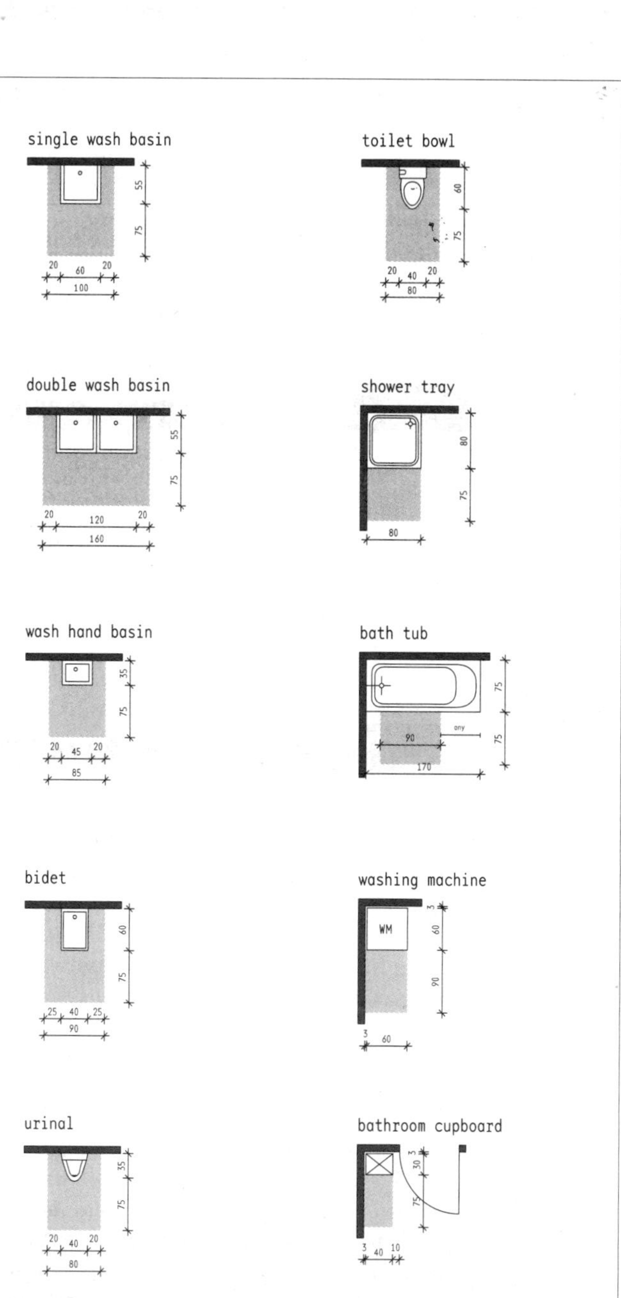

Fig.31:
Minimum movement areas in front of sanitary appliances

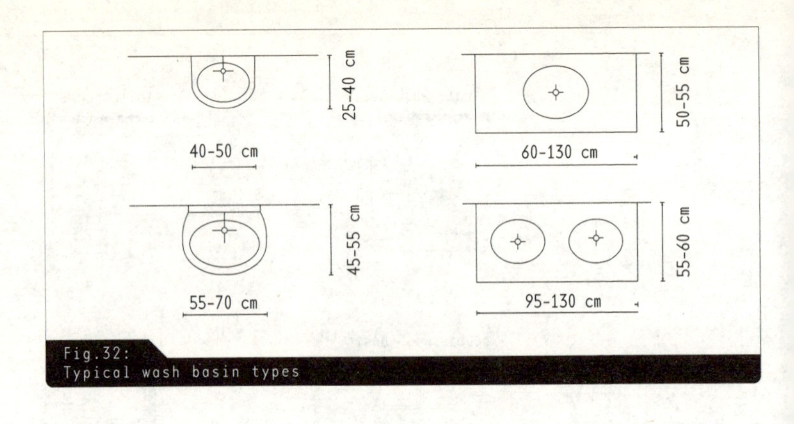

Fig.32:
Typical wash basin types

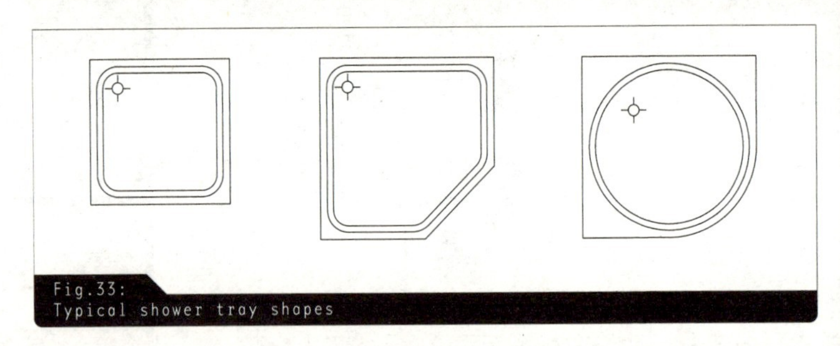
Fig.33:
Typical shower tray shapes

space than two separate wash basins; however, they must have a minimum width of 120 cm so that two people can use them unhindered at the same time. › Fig. 32

The corner valves for the hot and cold water pipes are installed under the wash basin and are used to turn off the drinking water to the fittings if repairs are needed. It is also possible to conceal waste and drinking water pipes and traps › Chapter Waste water, Waste water pipework in buildings in cupboards or behind panels under wash basins. Wash basins are also often built into specially designed bathroom furniture. This allows better use of the space in a bathroom and provides a more aesthetic way of concealing pipework.

Shower installations

Shower trays are normally manufactured from enameled cast iron, enameled steel plate or acrylic. Various shapes of shower tray are available. They could be anything from rectangular or square to circular or semicircular. › Fig. 33 The standard square shower tray is 80 × 80 cm in plan with a depth of about 15 to 30 cm; larger dimensions and lower entry upstands offer more space and are easier to use. To reduce the step up into the shower, the recess in the floor structure must be made deeper to allow

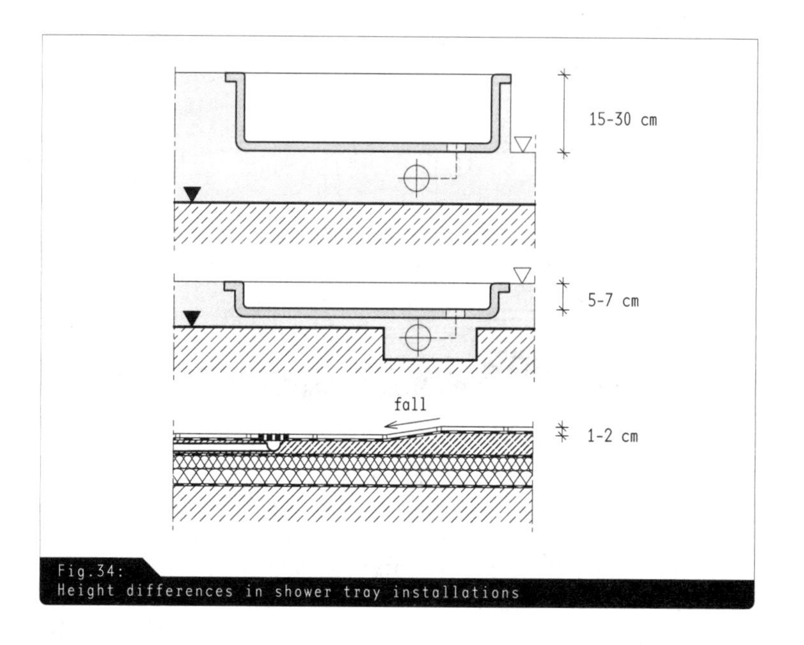

Fig.34:
Height differences in shower tray installations

for the waste water connection, which normally fits conveniently under the shower tray. › Fig. 34 top and middle

Some bathroom designs dispense with the shower tray and have the shower floor at the same height as the bathroom floor. › Fig. 34 bottom This offers not only design and cleaning advantages, but is also part of a barrier-free bathroom. With these wet-room showers, the drainage outlet is at floor level. It is therefore necessary to have efficient waterproofing and edge seals, and a higher floor level to provide the required fall in the waste pipe. ›

\\Note:
The waterproofing and seals may consist of waterproofing membranes and sealing tape, or other waterproofing materials, which are spread on the substrate using the thin bed process. They should extend at least 15 cm above the top of the floor covering and brought up even higher in the indirect spray zone of the shower–even if the shower is in the bath. They must also extend at least 20 cm above the shower head on the walls.

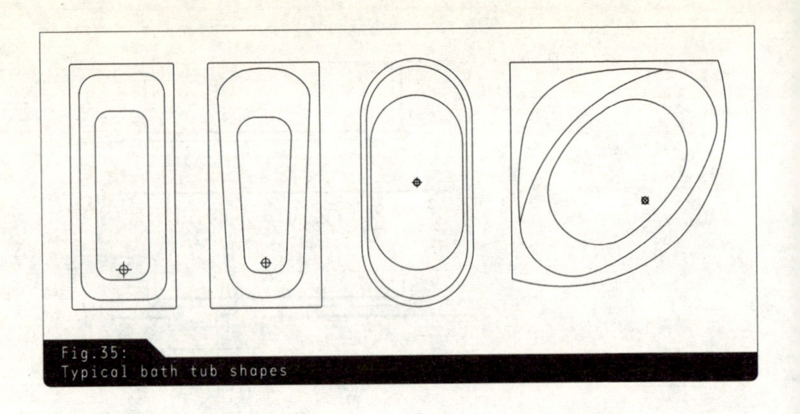
Fig.35:
Typical bath tub shapes

Bath tubs

Bath tubs may be freestanding in the room, placed along one wall and clad in subsequently tiled boards, or set in a preformed foamed plastic bath panel unit. They are manufactured from enameled cast iron, steel plate or acrylic, and are normally 170 to 200 cm long, 75 to 80 cm wide and 50 to 65 cm high. If the entry height needs to be lower, the bath must be recessed into the floor to allow the waste pipe connection to be made. The bath can be made flush with the surrounding floor, for example by building a raised floor plinth around it, as otherwise the floor must be considerably lowered in the area of the bath tub.

The voids between the bath tub and cladding are generally filled with insulation. The gap between the wall and the edge of the bath tub is made watertight with a flexible seal. A clear area of about 90×75 cm should be provided in front of the long side of the bath tub to allow unobstructed entry.

Lavatory pans

Lavatory pans can be wall-mounted or stand on the floor. › Fig. 36 The wall-mounted variety simplifies floor cleaning and is usually fixed to special framing members in the false wall construction so that its height can be adjusted. The shape of the lavatory pan depends on the type of flushing process. While older buildings mainly had shallow-flush pans installed, they are now increasingly being replaced by the less noisy deep-flush pans. › Fig. 37

WC flushing cisterns

Flushing cisterns range from false wall installations or exposed pressure chamber cisterns up to wall-mounted, inbuilt or close-coupled cisterns. › Fig. 38 They can be installed at various heights above the pan. While high-level cisterns, which operate with considerable noise, are common in older buildings, new buildings have low-level cisterns or cisterns built into the walls, which are substantially quieter. Pressure cisterns use the pressure in the drinking water pipe and therefore do not require a conventional box-like cistern. The self-closing valve provides flushing water

Fig.36:
Wall-hung and floor-standing WCs

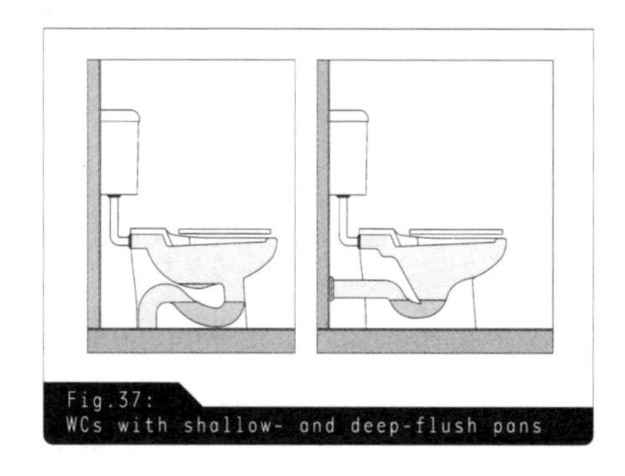

Fig.37:
WCs with shallow- and deep-flush pans

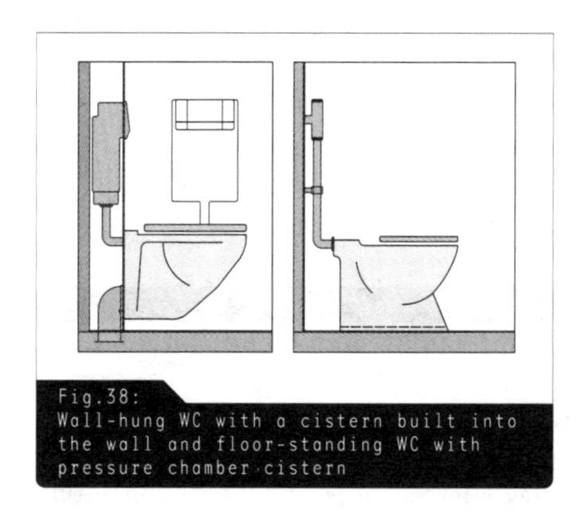

Fig.38:
Wall-hung WC with a cistern built into the wall and floor-standing WC with pressure chamber cistern

only as long as it is needed, i.e. as long as the flush lever is pressed. Conventional cisterns, on the other hand, are automatically refilled after every flush.

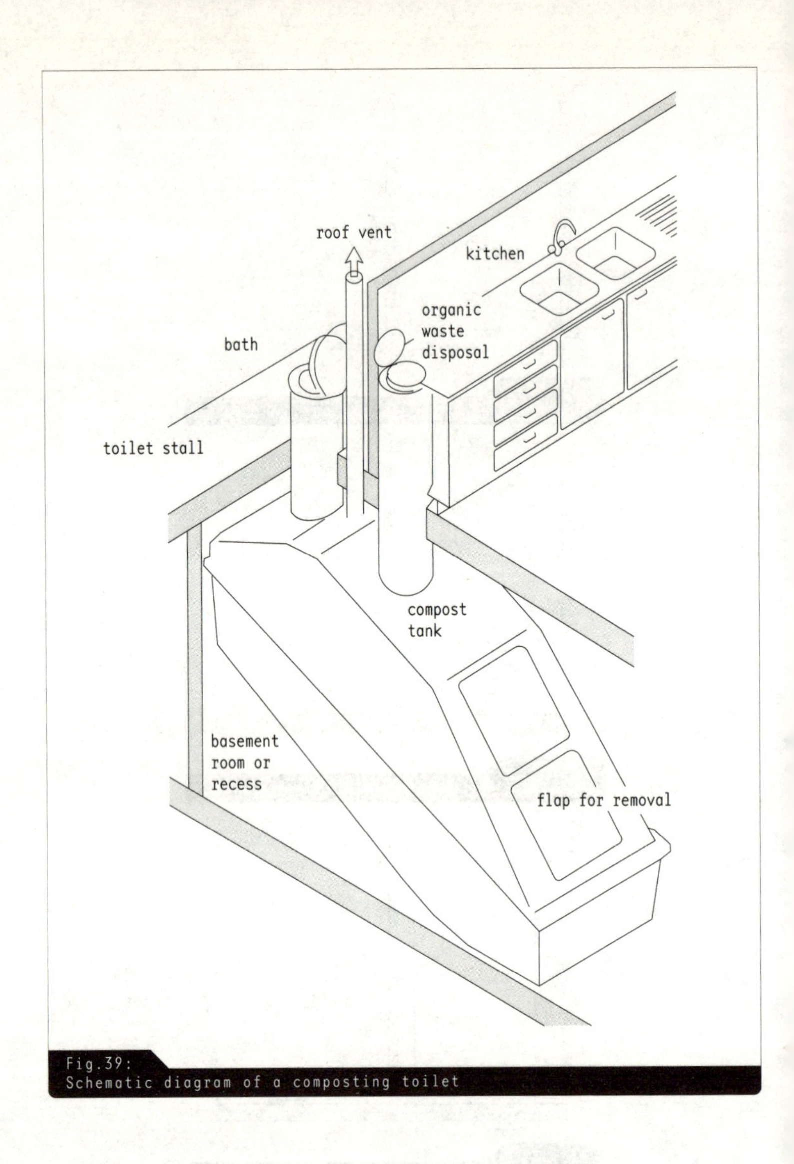

Fig.39:
Schematic diagram of a composting toilet

Water-saving toilet systems

More than one third of the daily water demand per person, approximately 35 to 45 l, is used for toilet flushing. Reducing consumption of water for flushing offers great potential for saving drinking water. Older cisterns use between 9 and 14 l water per flush, while modern toilet systems require about 6 l. The quantity of flushing water can be set by adjusting the filling height in the cistern. It should also be possible to interrupt the flushing process by pressing the flushing button a second time (water-saving button). To reduce the quantity of water used per flush to as low as 3 l requires the cistern to be connected to a special type of lavatory pan, as otherwise noise may become a nuisance.

Vacuum toilets

Vacuum toilets need only 1.2 l water per flush. They have a history of use on board modern high-speed trains and ships. In residential buildings a pump sucks out the contents of the toilet and delivers it into a ventilated waste water tank. From there it is transported by another pump to the public drainage system. The lower drinking water consumption achieved using the vacuum process could result in high savings of waste water disposal charges. The smaller pipe cross sections make installation of vacuum toilets problem-free. However, the flushing process itself is considerably noisier than other types of WC systems.

Composting toilets

Composting toilets do not use any water for flushing and therefore produce no waste water. They are used for ecological reasons or because the building is not connected to the public sewer. These toilets consist of a tank with one connected shaft for organic kitchen waste and one for toilet waste. › Fig. 39 Constant underpressure in the composting tank means that no odors can escape into the rooms. The decomposition of the material in the tanks over a period of months is initiated by air flowing through it. The nutrients resulting from this composting process can be used for soil improvement and plant food in the garden.

Fittings

The term fittings covers all shut-off devices fitted to a drinking water supply system such as stopcocks, gate valves or stop valves, as well as sanitary fittings on wash basins and showers, for example. Shut-off valves block or allow flow along lengths of pipework; they differ in the way they close off the pipe. Stop valves divide up a drinking water system into logical sections to allow parts of the system to be isolated and individual components replaced. Stop valves are therefore fitted before and after the water meter, filter and pressure reducer › Fig. 40 or at the lowest point of each riser and pipe supplying each story. In this way components can be replaced without having to shut down the whole pipe network. Stop valves are also installed at WC cisterns and under wash basins.

Wall- and surface-mounted faucets

Tap fittings in sanitary rooms are available as wall- or surface-mounted. › Fig. 41 Wall-mounted faucets, which are mainly installed at bath tubs or showers, are fitted directly on to the wall concealing the water supply pipes, where a short connection piece is used to connect to the drinking water pipework. Surface-mounted models are attached directly to the top of the wash basin or sink and connected to the drinking water supply by means of corner stop valves. The type of tap fitting depends on the intended use; kitchen faucets, for example, have a longer spout than bathroom faucets.

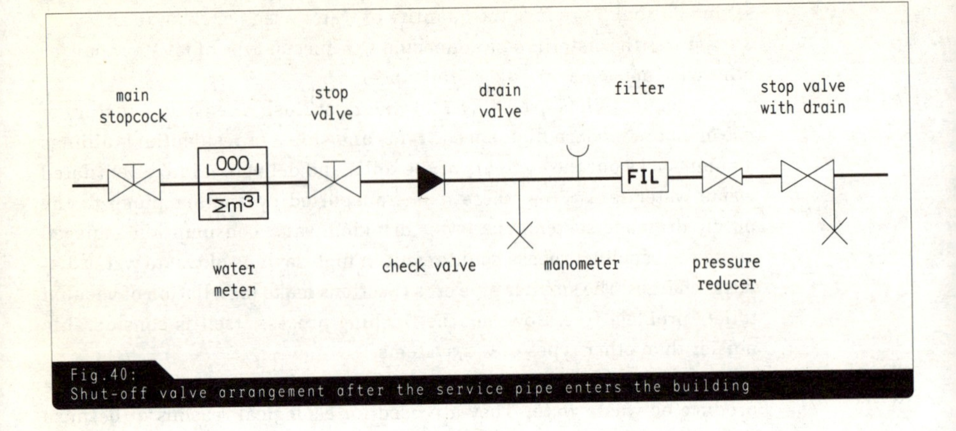

Fig.40:
Shut-off valve arrangement after the service pipe enters the building

Mixers

Conventional mixers consist of two rotating handles positioned close to one another; with the right and left handles controlling the cold and hot water flow respectively. The user controls the water temperature manually by adjusting the flow of each. More practical are the single lever mixers, which control the water temperature by rotation and the flow by an up or down movement of the lever.

Contactless wash basin faucets

Contactless wash basin faucets are often installed in public toilets for reasons of hygiene. The movement of a hand in front of an infrared sensor opens the valve. Some electronically controlled versions just require a hand to go near the faucet to activate the water flow. A water flow regulator ensures that a constant amount of water issues from the tap. Contactless wash basin faucets require a source of electricity to work. This may be provided by batteries or an external power supply.

Thermostatic faucets

Thermostatic faucets allow the temperature of the water to be preset by a rotating handle, so that the water temperature remains constant even if the rate of flow alters. The faucet mixes hot and cold water at the correct ratio to achieve the set temperature. > Fig. 42

\\Note:
All faucets and sanitary appliances should be aligned with the tile grid in the room to produce a pleasing appearance. Faucets should be positioned at a tile joint, a joint intersection, or in the center of a tile.

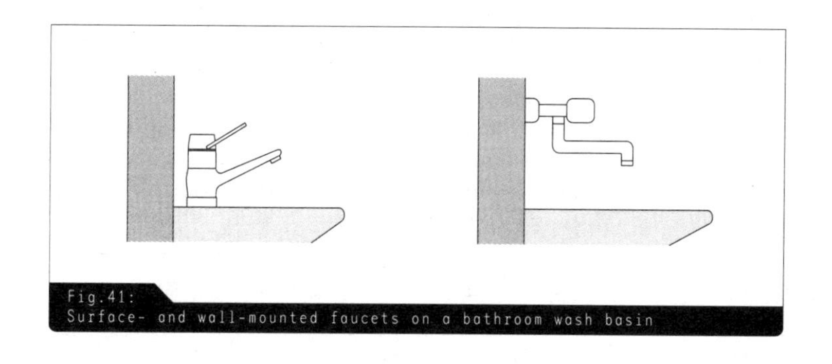

Fig.41:
Surface- and wall-mounted faucets on a bathroom wash basin

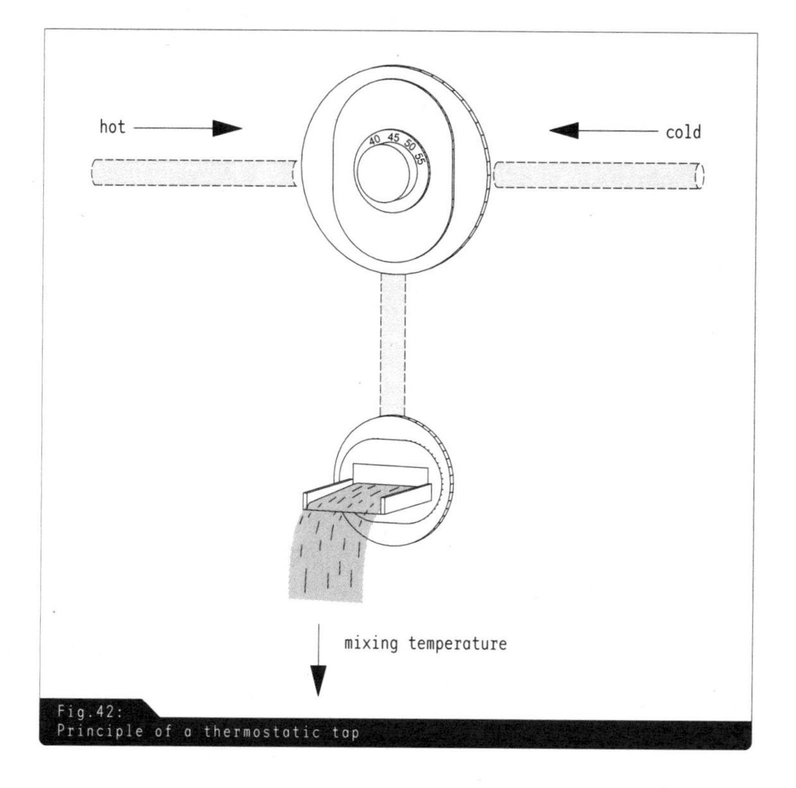

Fig.42:
Principle of a thermostatic tap

Barrier-free sanitary rooms

Barrier-free sanitary rooms have to satisfy special conditions. They should be fitted out in such a way that the occupant can use facilities in the room without help from another person. To achieve this there should be an adequate and barrier-free movement area of 120 × 120 cm in front of the wash basin, WC, shower and bath tub; for wheelchair use it would need to be at least 150×150 cm. Stepless shower trays level with the floor, wash basins that wheelchair users can drive under, and grab handles near all sanitary appliances make their use much easier. The door should have a clear opening width of at least 80 to 90 cm and should open outwards so as not to interfere with access to the sanitary appliances in the room. › Fig. 43

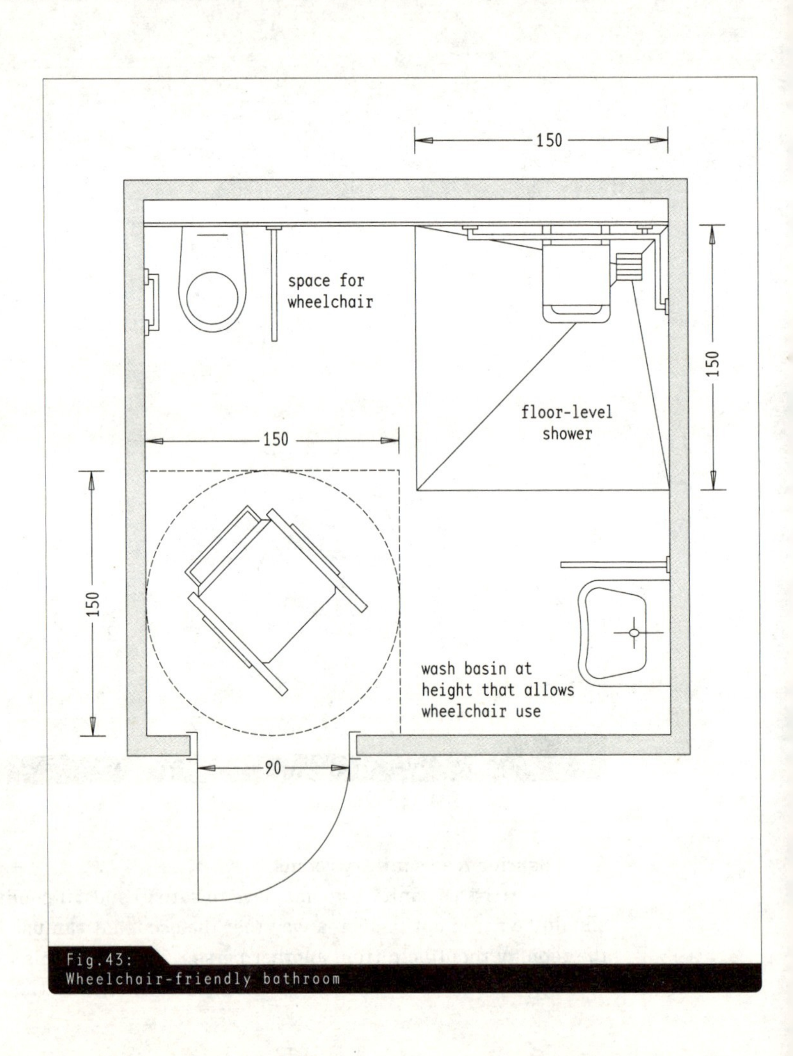

Fig.43:
Wheelchair-friendly bathroom

WASTE WATER

After it has been distributed through the building's network of pipes, drinking water is automatically changed into waste water when it flows out of the faucets at draw-off points into wash basins, showers or bath tubs, even though it might be absolutely unused and clean. The description "drinking water" ceases to apply as soon as the water enters the waste water pipework.

The term "waste water" generally covers not only water that has become contaminated by domestic, commercial or industrial use but also relatively clean precipitation water (rain). Waste water is usually contaminated by solid particles, bacteria or chemicals, and must therefore be thoroughly treated before it can be fed back into natural bodies of water. This process is normally carried out at the public waste water treatment works.

Heavily contaminated domestic waste water from toilets and dishwashers that contains fecal and putrefactive substances is known as black water. Less heavily contaminated waste water from wash basins, showers and bath tubs is called gray water and contains only about one third the contaminants as black water. This difference is immaterial for the normal process of waste water treatment at the sewage works is concerned, as the plants are generally designed to purify black water. This difference is, however, very important for natural waste water treatment processes, as some facilities can only treat gray water.

Increasing awareness of our environment has shifted the focus, for several years now, to protecting the purity of groundwater, rivers and lakes. The biological self-cleaning process does not work above a certain level of contamination, so that waste water treatment methods of purifying heavily polluted industrial and domestic water are of the utmost importance in counteracting ecological damage. › **Chapter Waste water, Methods of waste water treatment**

\\Example:
A four-person household in Germany introduces about 100 kg of detergent into the sewage system. The development of more environmentally friendly detergents has lowered the pollutant load, but has only slightly reduced the problem.

But before treatment can take place, the used cold and hot water must be transported from sanitary rooms in buildings and fed into the public drainage system. This takes place through the pipe network described in the section below.

WASTE WATER PIPEWORK IN BUILDINGS

Waste water pipes are considerably larger than drinking water pipes and have the task of taking rain and dirty water away from buildings and conducting it into the sewers. A branched network of pipes of different sizes is required ensure that the sanitary appliances continue to work without problems. Buildings are generally designed to be drained by gravity, so all pipes are either vertical or are installed at a fall of at least 2% to transport the waste water down and out of the building. It is important that the waste water does not back up in the system. ›

Flood level

The flood level is the maximum possible level up to which waste water may rise at a particular location in a drainage system. › Fig. 44 Normally the top of the road surface or the top of the curb at the connection point can be taken to be the flood level, unless indicated otherwise by the local flood prevention authority. This is the limit to which the water will rise in the event of a flood, and therefore it cannot back up any higher inside the building. A flood is most likely to occur during periods of heavy rain. Combined drainage systems are particularly at risk because they carry waste water and rainwater together. › Chapter Waste water, Methods of waste water treatment Flooding also occurs in separate drainage systems, for example if pipes become blocked.

In connected sanitary appliances that are below the official flood level in basements, there is always a risk that waste water from the drainage system will enter the building and cause serious damage. For this reason, each connection point must be protected with an anti-flooding valve

\\ Note:
If waste water backs up it is possible for the waste water in the public sewer to enter the building's pipework system connected to it, based on the principle of communicating vessels: if vessels or pipes that contain liquid and are open at the top are connected to one another, the level of the liquid will be at the same height in both-irrespective of their shape.

\\ Note:
European Standard EN 12056 applies to gravity drainage systems inside buildings. EN 752 applies to systems outside buildings. Both standards set out a general framework, which requires national annexes and allows regional departures.

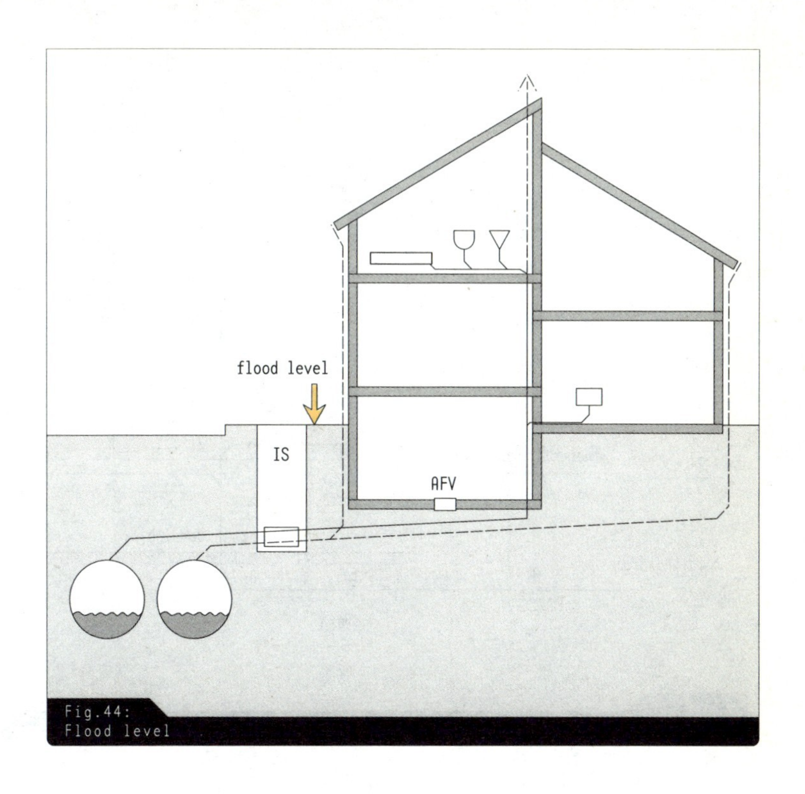

Fig. 44:
Flood level

or a waste water lifting plant. › **Chapter Waste water pipework in buildings, Protective measures, p. 60**

System components and pipe runs

Individual and common waste pipes

A pipework system is made up of many different components, which are connected together to conduct waste water into the public drainage system. › **Fig. 45** An individual waste pipe connects each sanitary appliance to a common waste, into which all the wastes in a sanitary room are brought together. › **Fig. 46** The common waste has a fall of 2% and takes the shortest route to the vertical stack, which in turn carries the waste water downwards in a uniform diameter pipe with as straight an alignment as possible. The fall in horizontal waste water pipes is necessary to ensure that the flowing water leaves behind no residues in the pipe. All pipes are normally connected into the discharge pipe by 45° bends in the direction of flow to prevent water building up at the connection point. Neighboring common waste pipes must have staggered connections into the vertical stack to prevent water in one from entering the other. The individual

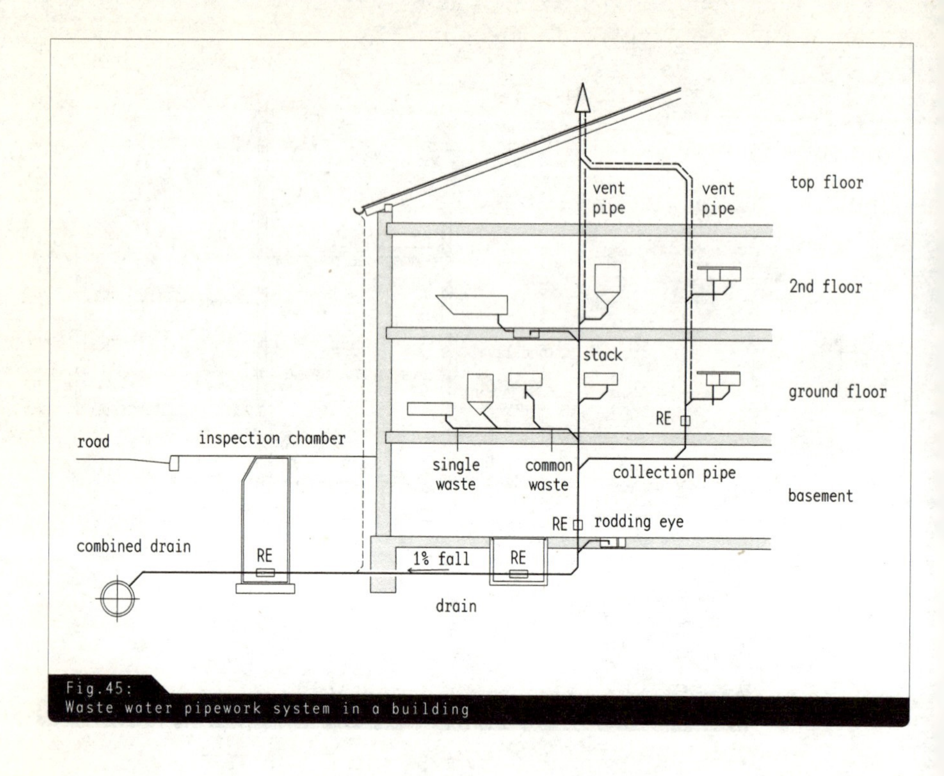

Fig.45:
Waste water pipework system in a building

pipework system components may be screwed together, solvent welded, or have push-fit connections.

Stack, drainage and vent pipes

The vertical stack pipe normally discharges below the level of the building's floor slab into a drainage pipe, which leads to the public drainage system or sewer. The stack must be vented to prevent backflow into sanitary appliances. This backflow is caused by underpressure in the stack resulting from the pressure differences that occur when the stack is suddenly used by several appliances at once. If the total length of stack is more than 4 m, it obviously runs through more than one story, and must therefore vent into the open air above the sanitary appliances in the top story and roof level without any reduction in cross section. Alternatively it could be fitted with a venting valve below roof level specially designed for waste water pipes. If the venting pipe is led through the roof to the open air, it must be at least 2 m away from a dormer window or roof window, or must project 1 m above the highest point of these features, so that no unpleasant odors can enter the building from the waste pipe.

At the lowest point of the stack, a drain sufficiently deep underground so as to be frost-free transports the waste water from the building into the

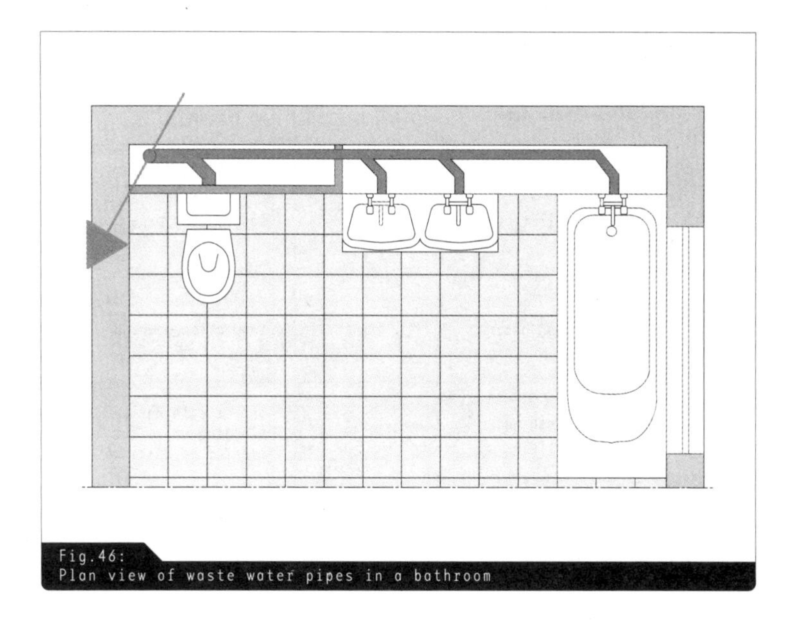

Fig. 46:
Plan view of waste water pipes in a bathroom

connection drain, which is connected directly into the public sewer. If an ordinary buried drain is out of the question because the building has a basement and the public drainage system is too high to allow a normal connection, a collection drain can be laid to a fall below the basement floor slab.

Access for cleaning

Only junctions with angles of up to 45° are permitted in buried and collection drains so that waste water can flow smoothly away. In addition, rodding eyes or similar openings must be provided at least every 20 m to allow any length of pipe to be unblocked and cleaned out without excessive effort. In vertical stacks, there must be a suitable cleaning opening at the lowest point, because this is where a blockage is likely to occur first.

Informative signage

The position under the road of drains in the public waste water drainage system is indicated by sign plates usually fixed to building walls or marker posts. The numbers give the direction and distance of connection drains.

Drawing symbols

Various symbols are used in plan and sectional views of planned waste water systems to improve the readability of the drawings and to show the numbers and arrangement of the connected sanitary appliances. › Fig. 47 In a similar way to that described in the chapter on water supply, the drawings should show the true position of objects in plan marked with the appropriate symbol in conjunction with the pipework system. The pipework system, including the pipe layout and sanitary appliances, is shown in a schematic sectional view as if the wash basins, showers, bath tubs or

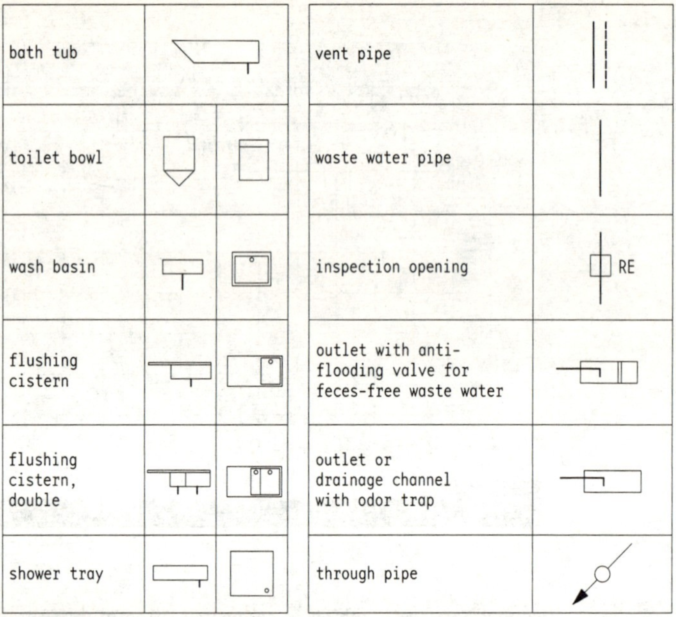

bath tub			vent pipe	
toilet bowl			waste water pipe	
wash basin			inspection opening	RE
flushing cistern			outlet with anti-flooding valve for feces-free waste water	
flushing cistern, double			outlet or drainage channel with odor trap	
shower tray			through pipe	

Fig.47:
Representation and explanation of commonly used symbols

WCs were adjacent to one another and all connected to a single common waste. › Fig. 45 The pipes are shown with the 45° bends mentioned above, appropriately arranged for the actual direction of flow.

Calculation of pipe sizes

The cross section of the pipes depends on the type and number of connected sanitary appliances and the water demand, which in turn depends on the desired level of comfort and convenience in the building. Each appliance that generates waste water has a connection value (DU) and a minimum required pipe cross section. › Tab. 6 The estimated waste water outflow (Q_{ww}) in liters per second (l/s) of an appliance is the most important parameter in calculating the required nominal pipe size in waste water systems. The drainage index (K) is a measure of the frequency with which a waste water appliance is used. Hence the design of the system will vary with the type and use of the building. For example, sanitary installations in schools or public buildings will be much more frequently used than those in residential buildings.

Tab.6:
Typical connection values of sanitary appliances and pipe diameters of single wastes

Waste water appliance	**Connection value (*DU*)**	**Pipe diameter**
Wash basin	0.5	DN 40
Shower with plugged outlet	0.8	DN 50
Shower with unplugged outlet	0.6	DN 50
Bath tub	0.8	DN 50
WC with 6-liter cistern	2.0	DN 100
WC with 4- to 5-liter cistern	1.8	DN 80 to DN 100

The sum of the connection values of each appliance is used to calculate the pipe cross section of the common waste, stack or drain pipe. Therefore the required size of the drain is calculated from the sum of the connection values of all the waste water generating appliances connecting into it. A horizontal waste water pipe to which a WC is connected, for example, usually requires a nominal pipe size of at least DN 100; which means a pipe with an internal diameter of 100 mm. The vertical stack must therefore also have a diameter of at least DN 100. For a series of connections into a pipe, the total connection value can be calculated using the following formula:

$$Q_{ww} = K \times \sqrt{\Sigma\ (DU)} \text{ in l/s.}$$

Q_{ww} = Quantity of waste water (waste water outflow):
DU = Design units (connection value);
K = dimensionless drainage index, which represents the frequency of use (in residential buildings 0.5; in schools, restaurants, hotels 0.7; in public buildings with frequent usage 1.0).

Materials

Waste water pipes can be made from vitrified clay, cast iron, steel, fiber cement or plastic, while rainwater downpipes may be manufactured from lead. Vitrified clay pipes are generally used for buried drains, as they are resistant to load. Cast iron and fiber cement pipes are suitable for all types of building and land drainage systems. Their high mass per unit

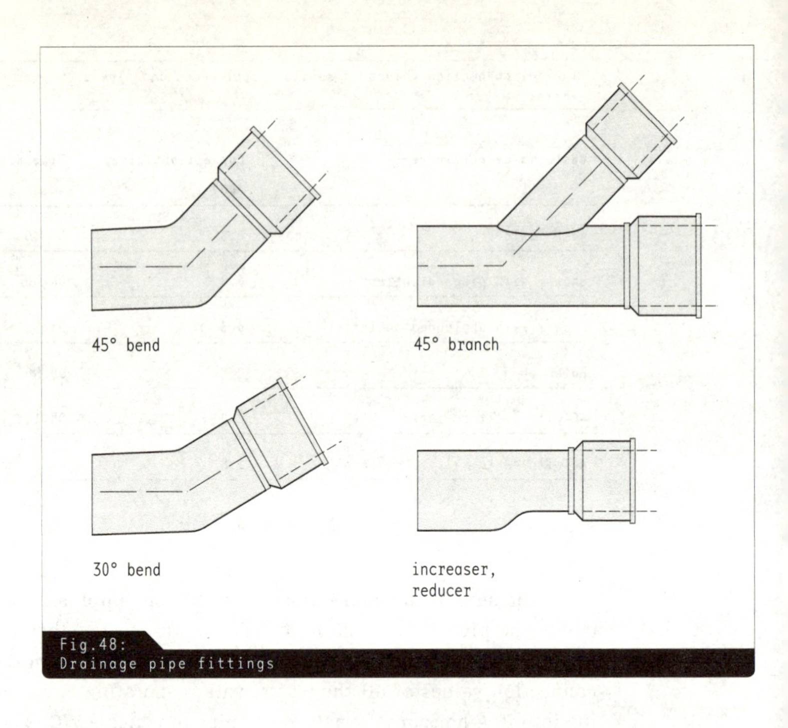

Fig.48:
Drainage pipe fittings

area makes them particularly useful for attenuating the noise of flowing waste water. Steel or stainless steel pipes are used where the waste water they carry is corrosive, which might be the case in laboratories, for example. The most economic material is plastic. Low weight and corrosion resistance means plastic pipes are primarily used in residential buildings, with higher-quality plastic pipes also finding use in industrial and commercial buildings. The plastic used for all pipe components must be heat-resistant.

All materials are produced in short standard lengths and connected to one another by push-fit sleeves, threaded or sealed connections or, in the case of rainwater downpipes, crimped or soldered. They can be obtained as bends, branches, increasers and in other shapes. › Fig. 48

Protective measures

Odor traps

Traps ensure that no unpleasant odors can escape from waste water pipes into rooms. They are fitted below the outlet of each sanitary appliance. Traps are available in various forms, but they all work on a similar principle to the pipe odor trap, which is the most popular due to its excellent flow characteristics. It consists of a curved piece of pipe of at least 30–45 mm diameter, which retains some water in its bend. › Fig. 49, left The

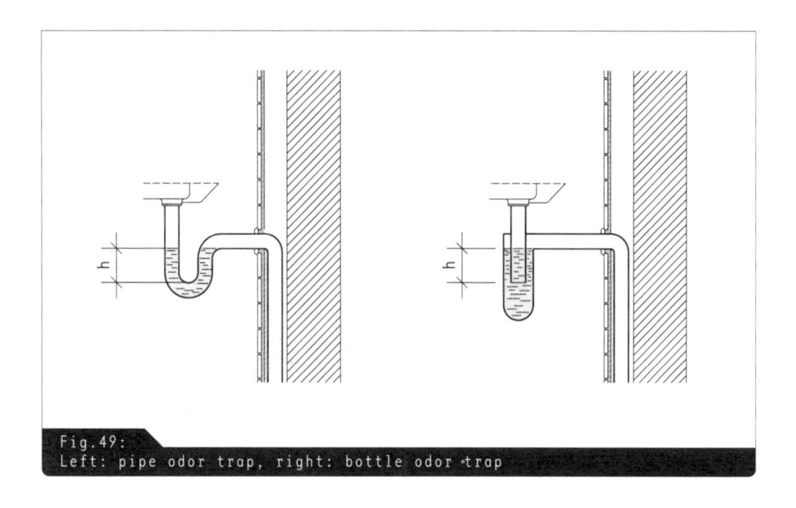

Fig.49:
Left: pipe odor trap, right: bottle odor trap

bottle trap is more prone to blocking and therefore is less popular. › Fig. 49, right The standing water in double traps prevents odors from escaping out of the pipe into the room.

Floor outlets

It is prudent to build floor waste water outlets in bathrooms in residential properties where washing machines or floor level showers are installed. They are specified for use in public buildings or swimming pools. They may be manufactured from cast iron, stainless steel, brass or plastic; they can be installed in the floor structure and require the least possible installation depth. When these floor outlets are installed, the floor must have a slight fall of 1.5% towards the outlet and be sealed so as to be waterproof. › Fig. 50 As the floor outlet is often in the middle of the room, it is not always an easy task to lead the connected waste pipe to the nearest stack. These outlets are usually only connected to DN 50 or perhaps DN 70 waste pipes, but they must be laid with the normal 2% fall.

Anti-flooding valves

As discussed earlier with reference to flood level, all waste water appliances that are below the flood level must have tightly sealing anti-flooding valves to prevent the backflow of foul water into the building. This may happen, in particular with combined drainage systems, when heavy rain takes the flow in the public drainage system to its limit. A high water level in the public drains can cause waste water from deep connection pipes to emerge from sanitary appliances. › Fig. 51

Anti-flooding valves usually consist of a motor-driven shut-off valve, a pneumatic gate valve, or an automatic or manually actuated stop valve (emergency shut-off). However, all the waste water pipes in the stories above flood level must not discharge through the anti-flooding valve, but

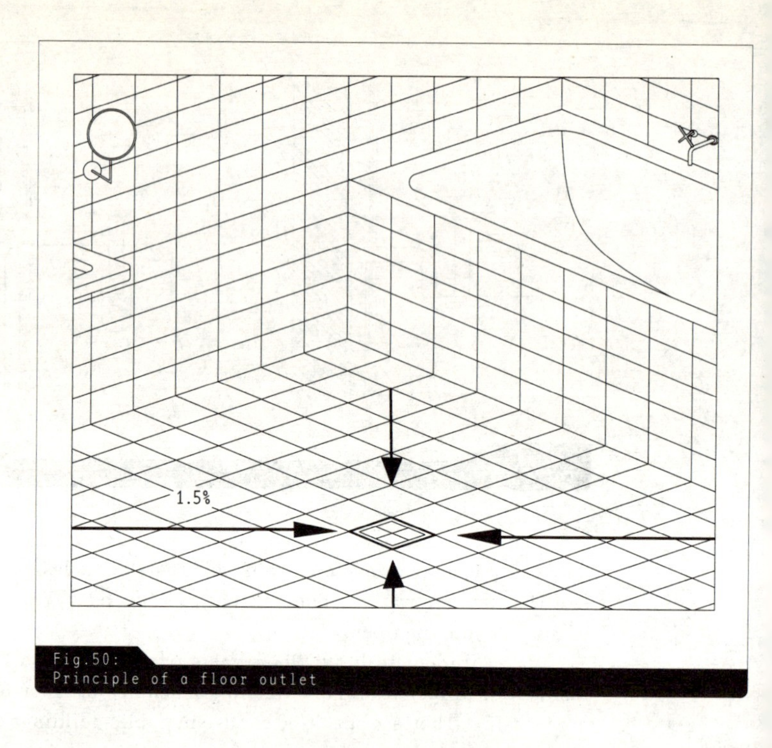

Fig.50:
Principle of a floor outlet

must be connected downstream of it, as the building could otherwise be flooded by its own waste water.

Waste water lifting plants

Waste water appliances situated below the flood level that cannot be connected with an adequate fall to the public drainage system because they are too deep underground have to be drained by a waste water lifting plant. This collects the waste water, which may or may not contain fecal matter, in a tank and delivers it by means of a pump and pressure pipes, through an anti-backflow riser with its highest point above the flood level, into the public drainage system. › Fig. 52 The height of the anti-backflow riser ensures that there is no way for the waste water to flow back into the building. The waste water is then taken in a drain connected to the plant's discharge pipe at the normal fall into the public drainage system.

METHODS OF WASTE WATER TREATMENT

Separate and combined drainage systems

Waste water is discharged into either a combined or a separate drainage system. In a combined system, domestic and industrial waste is led together with rainwater into the drainage system; in a separate

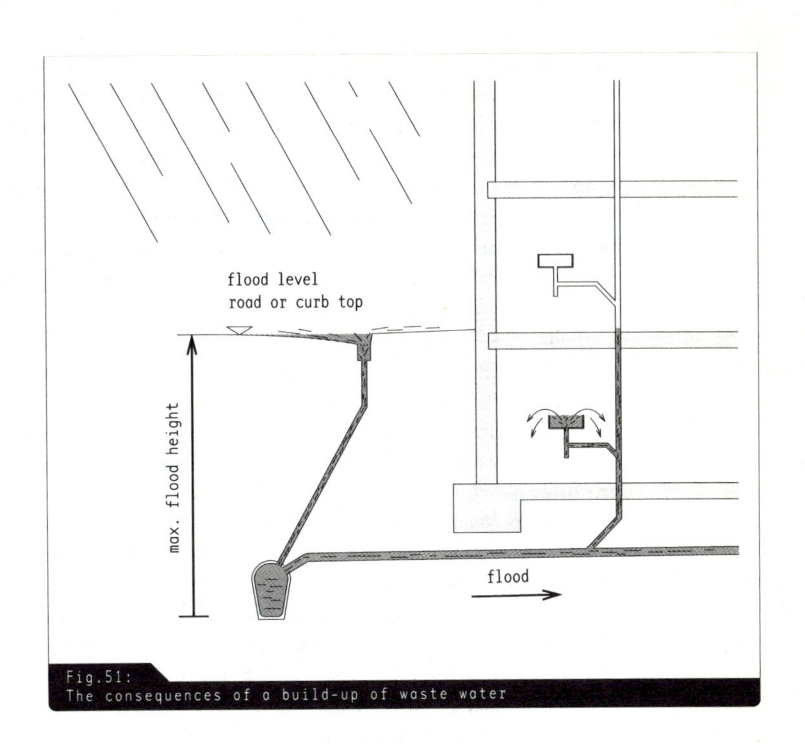

Fig.51:
The consequences of a build-up of waste water

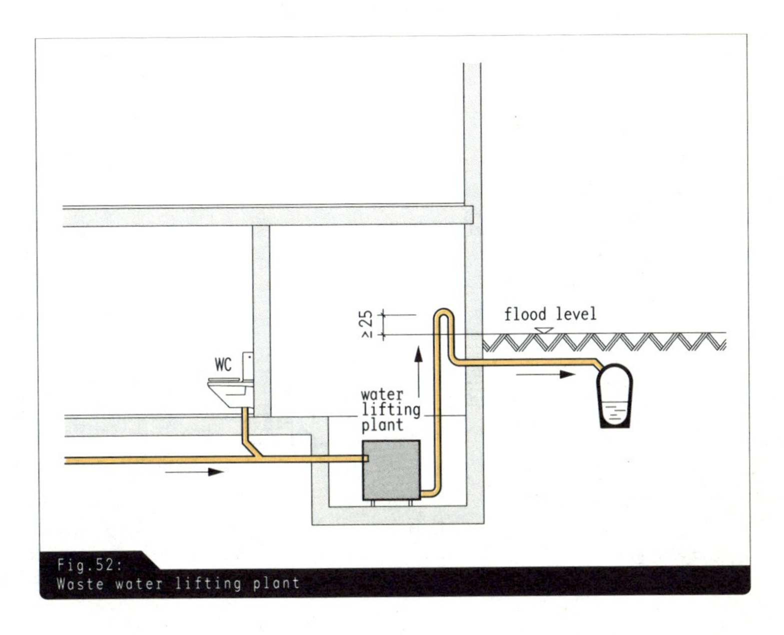

Fig.52:
Waste water lifting plant

system, rainwater flows directly into open water or watercourses, often referred to as outfalls, and only the contaminated water enters the public drainage system. › Fig. 53 Within the area of the building and in the design of the drainage pipework, rainwater is now assumed to be dealt with separately from waste water, even if the public drainage system outside below the road surface is a combined system, as many countries are making plans to move to separate systems at some time in the future.

The reason for this is that precipitation water only turns into dirty water when it is mixed with ordinary foul water in combined drainage systems. As a result the volume of waste water increases greatly, which makes changing to a separate system worthwhile to reduce the foul water load on the drainage system and the costs of waste water treatment. The reduction in waste water volumes and the maintenance of natural groundwater levels are two good reasons why rainwater should be drained away if possible within the building curtilage, or be allowed to soak away locally.

Cleaning waste water

Municipal waste water treatment plants first remove the coarse particles from the waste water, and then clean it biologically to kill bacteria and treat it chemically to remove phosphates, heavy metals and nitrates. › Fig. 54 After cleaning, municipal waste water is discharged through outfalls into natural watercourses. But in spite of complex, costly water treatment systems, excessive amounts of plant nutrients and pollutants from treated waste water are entering natural watercourses, where they stimulate increased vegetation growth.

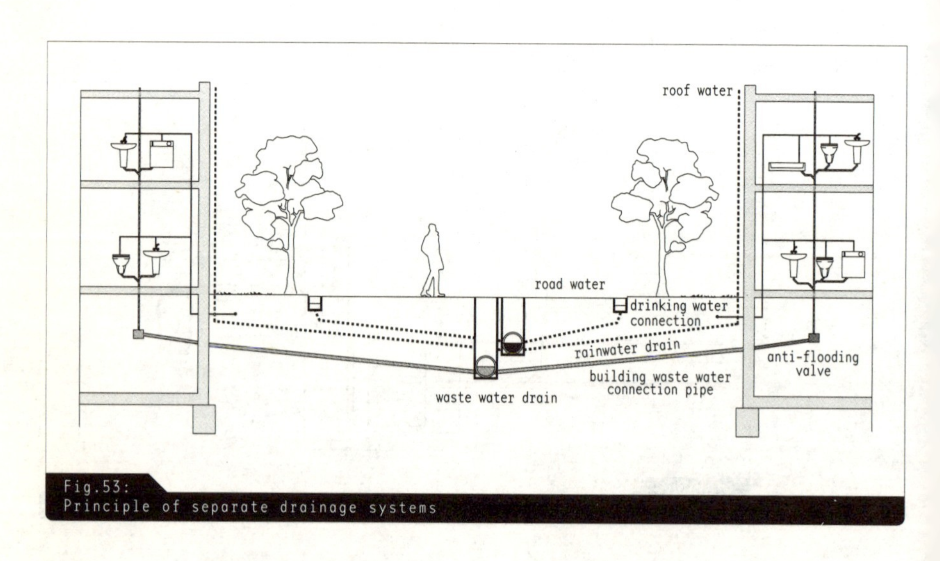

Fig.53:
Principle of separate drainage systems

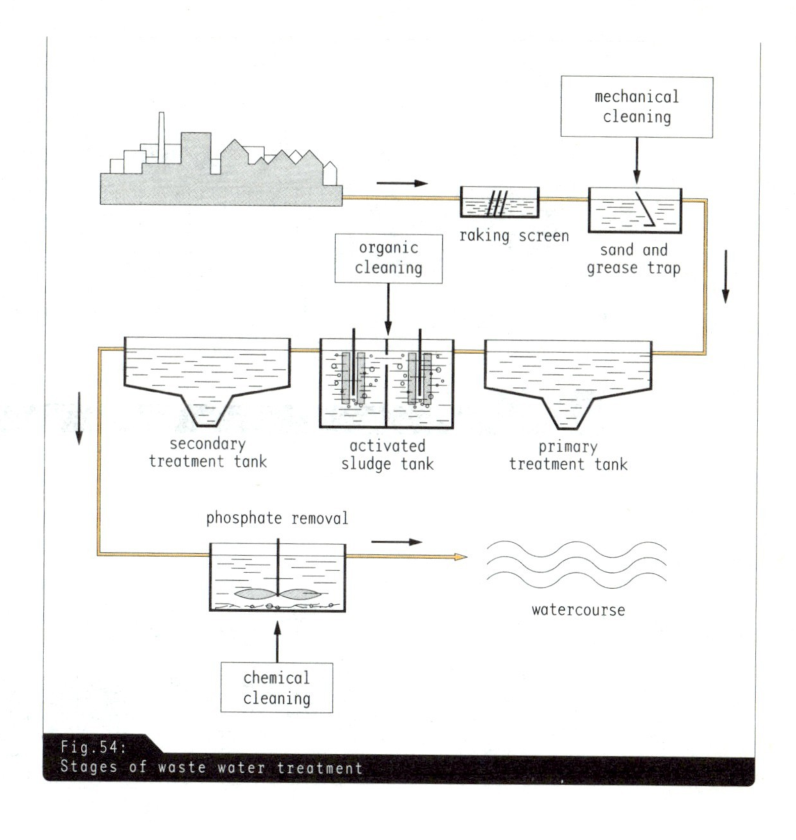

Fig.54:
Stages of waste water treatment

Natural waste water treatment systems

The idea of cleaning waste water using natural and less cost-intensive methods is not new. Water treatment facilities that work naturally are usually localized, small-scale disposal systems. Similar arrangements have been in use in rural areas for years, where the distance to a public drainage system may be too great to allow a cost-effective connection to be provided.

Increasing problems with the quality and costs of waste water disposal have meant that decentralized, natural water treatment methods have once again been considered as options by environmentally conscious planners over recent years. Numerous ecological housing developments have nearby reed bed water treatment systems, which clean all the waste water generated, within the boundaries of the site. These systems relieve the public drainage system, while sharpening our ecological sense of the natural water cycle and returning the responsibility for it to the individual.

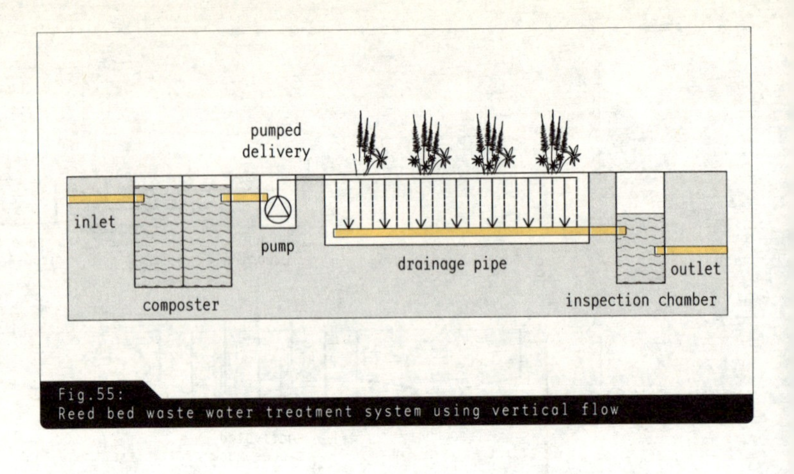

Fig.55:
Reed bed waste water treatment system using vertical flow

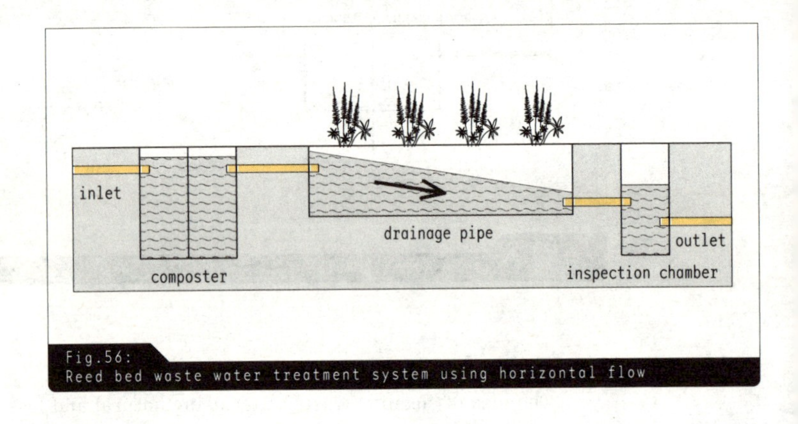

Fig.56:
Reed bed waste water treatment system using horizontal flow

Natural waste water processes have no need for the high energy and installation costs of conventional treatment works, but nevertheless produce excellent cleaning results. The actual waste water treatment process requires hardly any energy input from outside the site, but the systems themselves take up a lot of land. The design of the system largely depends upon the degree of contamination of the water to be treated.

Reed bed waste water treatment systems

Reed bed waste water treatment systems are the most common form of natural water treatment. They are generally based around waste water ponds with reeds growing in them and their cleaning effect is primarily produced by plant and animal microorganisms. Therefore it is not the plants themselves that purify the water; it is the microorganisms living in their roots that consume the nutrients in the waste water to produce the

Fig.57: Retention pond

cleaning effect. The bed of the waste water pond is usually a sand filter through which the water flows either vertically or horizontally. A composting chamber or three-chamber septic tank may be installed upstream of the beds to remove solid matter. A continuous flow of air through the chamber ensures that composting takes place.

First, the slightly soiled gray water from showers and wash basins or black water from toilets is carried along a separate pipework system by gravity, out of the building and through a buried composter, where the coarse particles are removed. From there the waste water is pumped out across the reed bed using the vertical flow principle, which takes up less area but the filter is further underground. › Fig. 55

Horizontal flow, on the other hand, involves a larger area of land but less depth. Here, the foul water flows out slowly across the system and is cleaned through the reed bed. › Fig. 56 The choice of system depends on the available space outside the building. Some waste water treatment methods use both flow types in series to achieve the best cleaning effect.

The reed bed is simply a sand and gravel filter bed. As most of the waste water seeps directly into the soil filter, the area does not look like open water but more like planted ground. The water from multistage systems can be passed through an inspection chamber, in which the water quality can be regularly tested before being taken to an outfall or a retention pond, where it could be used for recreation if the quality is adequate. › Fig. 57 The cleaned water could also be used as service water for flushing toilets. › Chapter Waste water, Uses of waste water

Natural waste water treatment processes require a great deal of space to be able to function properly, especially when used for cleaning black water. However, the aesthetically pleasing appearance and natural value of a reed bed are certainly more appealing than a conventional waste water treatment works. Reed bed waste water treatment systems should not be viewed just as an ecological alternative to conventional treatment processes. A natural treatment system is perfectly capable of handling all the waste water from buildings that are too far from the public drainage system.

DISPOSING OF RAINWATER

Rainwater is also part of the water cycle in buildings because it flows off roofs and impervious surfaces as dirty water and must be disposed of along with domestic waste water.

The solidly paved surfaces in urban areas prevent rainwater from simply seeping into the ground and becoming groundwater in the natural cycle, as would be ideal. Instead, it is conducted through pipes into the drainage system. › Fig. 58 During heavy rainfall the capacity of the drains is often inadequate, so that more and more foul water mixed with rainwater is flowing untreated into rivers and lakes.

Instead of rainwater being conducted rapidly into the drains, the modern approach is towards slower and more sophisticated systems of rainwater disposal that avoid mixing it with foul water. In choosing a suitable system, the designer must take into account the frequency and quantity of rainfall in the area, the character of the ground surface, and the height of the water table.

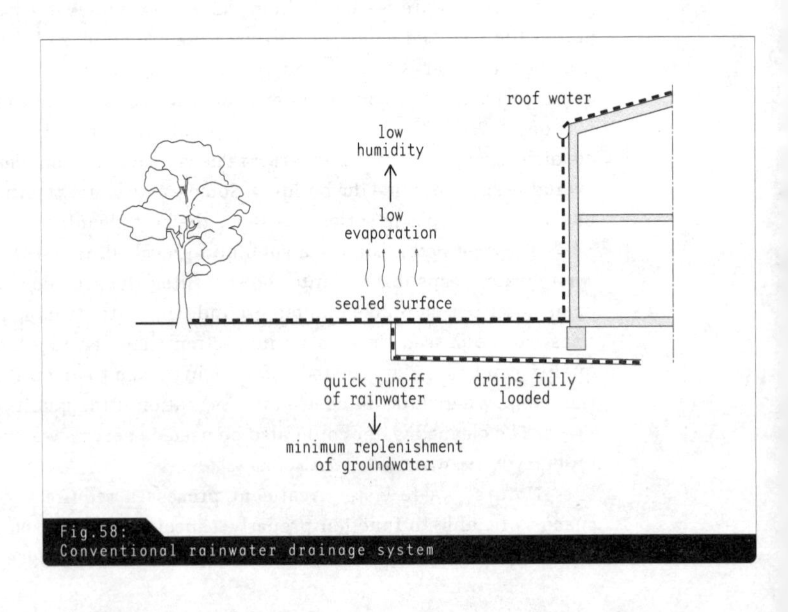

Fig.58:
Conventional rainwater drainage system

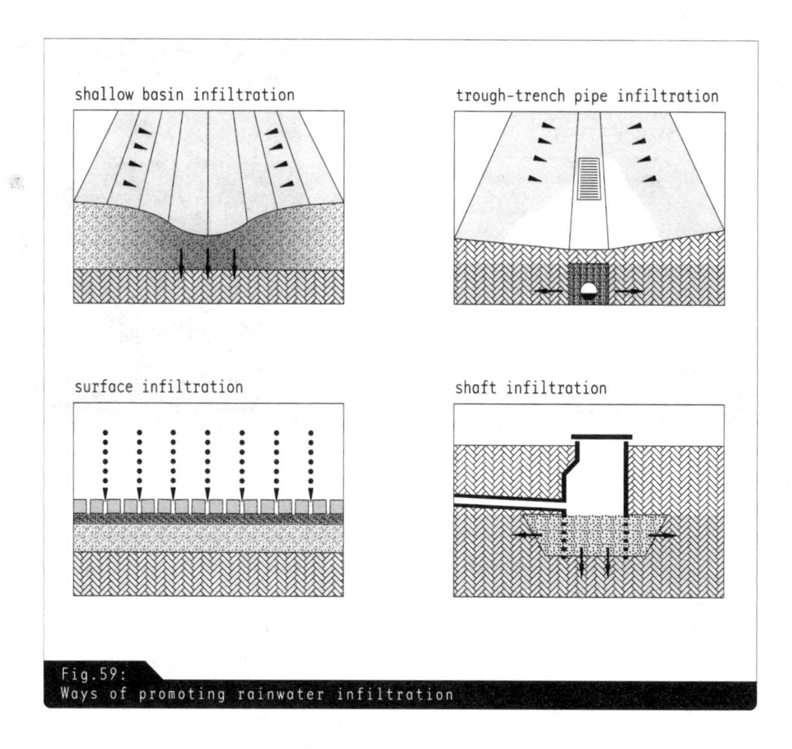

Fig.59:
Ways of promoting rainwater infiltration

Rainwater infiltration

To maintain the natural water cycle, ground surfaces that have not been built over, such as open ground, footpaths and squares, particularly in residential areas, should be designed to be as pervious to water as possible, for example as lawns or gravel. This can also be achieved by making built-upon surfaces less impervious so that rainwater can enter the ground in a natural way and contribute to a rise in the water table.

The character of the ground is critical to allowing rainwater infiltration. The more sandy a soil, the more pervious the ground is, and infiltration occurs naturally without any problems. If the soil is so loamy or clayey that rainwater is prevented from infiltrating, special measures may be required such as grass-lined basins, which store the water for a short time, trough-trench or pit infiltration systems, etc. › Fig. 59 These delay the runoff of rainwater and hold back the water in heavy rain, preventing the drains from becoming overloaded and reducing peak water levels. The urban climate benefits greatly from rainwater infiltration.

Costs can be saved in drainage by incorporating infiltration systems on the site. But costs may rise again due to the care and maintenance

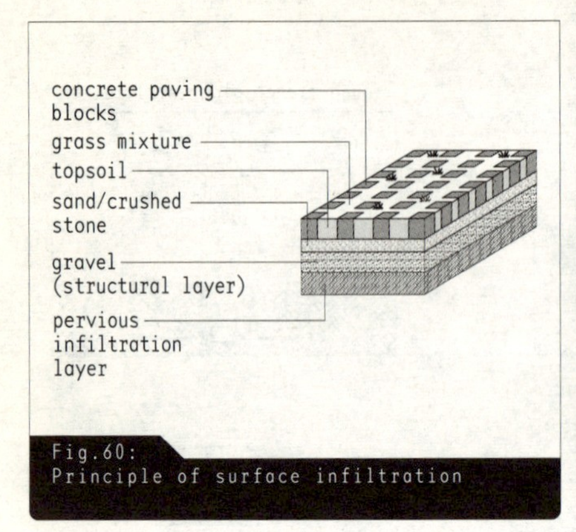

Fig.60:
Principle of surface infiltration

Fig.61:
Cellular grass paving for surface infiltration

required for more complex retention systems such as green roofs in combination with rainwater ponds or extensive infiltration systems—for example, grass-lined basins.

Surface infiltration

With surface infiltration, rainwater seeps into the ground without having to be temporarily stored. Cellular grass paving is one method of surface infiltration; › Figs. 60 and 61 another is water-pervious blockwork; and both are particularly useful for parking spaces, gardens, or little used vehicular accesses. A lawn or gravel footpath can also aid infiltration where the underlying soil properties are favorable. Rainwater undergoes initial cleaning through surface infiltration, even in the top layers of soil. This cleaning effect continues as the rainwater flows slowly through further soil layers until it enters the groundwater.

Shallow basin infiltration

Shallow basin or swale infiltration is a form of surface infiltration that delays rainwater runoff. An infiltration basin is a grass-covered hollow in which rainwater is stored for a few hours. › Fig. 62 During this time the water infiltrates slowly into the soil and eventually enters the groundwater.

Shallow basin infiltration requires less land than surface infiltration. With a depth of 30 cm stored water, the area can be estimated as about 10 to 20% of the area of the connected roof surfaces. The rainwater is cleaned as it passes through the various soil layers. Infiltration basins are inexpensive to construct and require little maintenance. Furthermore, they can be integrated as design elements into leisure facilities and green spaces.

Trough-trench systems

A trough-trench system combines two methods of infiltration: the grass-covered infiltration basin, and a gravel bed (trench) in which a drainage pipe is laid. The top, approx. 30 cm thick, layer of the basin

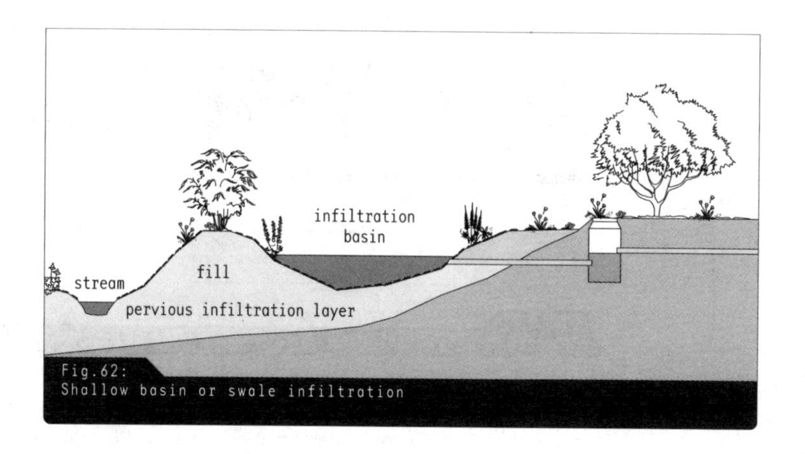

Fig.62:
Shallow basin or swale infiltration

acts as both store and filter for the rainwater. On leaving the basin, the rainwater is introduced at a single point or over an area into the trench, which is filled with coarse gravel and lined with a non-woven filter fabric. The rest of the rainwater is transported away slowly through the drainage pipe and is finally discharged at an outfall or into the public drainage system. On the way, the flow is dissipated through the porous pipe so that only a very small proportion of the rainwater arrives at the discharge point. Trough-trench infiltration is also suitable for handling high volumes of water in poorly draining ground conditions.

Rainwater retention

Retention is the holding back of rainwater. In large towns and cities, many millions of liters of waste water can be saved by holding back rainwater. Retention systems are intended to delay and reduce the direct flow of water into the drainage system. The water is delayed in most cases by green roofs or retention ponds. To an extent, and depending on the construction depth of the substrate, green roofs store precipitation water before releasing it, reduced by two thirds, into the drainage system. They improve the urban climate and in particular the local microclimate. The water evaporating from them can cool hot summer days and bind dust.

Green roofs

Green roofs may be extensive or intensive. While extensive green roofs have a substrate depth of 3 to 15 cm, an intensive green roof requires a substrate some 15 to 45 cm in depth. Both types of green roofs have a separating layer on top of the conventional roof membrane to prevent plant roots from destroying the roof construction. On top of this comes a drainage layer to remove the retained water, and finally the actual vegetation

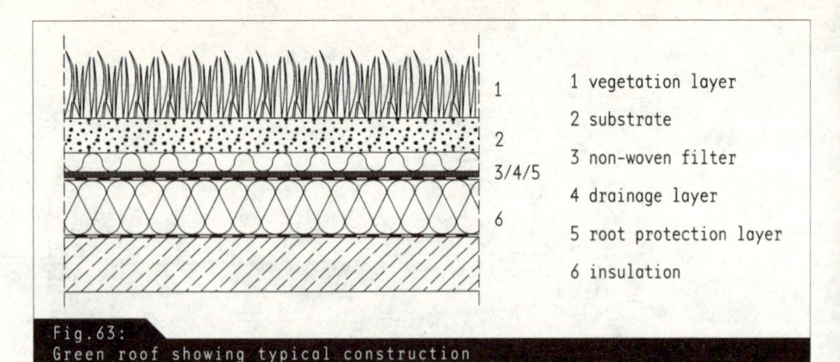

Fig. 63:
Green roof showing typical construction

layer. › Fig. 63 Roofs may be designed as warm, cold, or upside-down roofs, which differ from one another by the position and ventilation of the insulation layer.

Extensive green roofs with a substrate depth of between 3 and 7 cm are planted with mosses and succulents with minimal water and nutrient demand. A somewhat deeper soil construction is necessary for low or medium water demand plants such as grasses or more leafy vegetation. Intensive green roofs consist of faster-growing grasses, perennials or woody species. The greater the substrate depth, the more effectively the roof retains and evaporates water. On the other hand, planting costs and the load placed upon the roof also increase proportionately.

Any flat roof can be a green roof provided that the roof structure is suitable, adequately sealed, and able to support the extra load of the substrate. Roofs inclined at up to 15° require no special safety measures; for more steeply sloping roofs, precautions must be taken to prevent the soil from shearing or sliding off the roof.

Plants and soil layers on green roofs mechanically clean the water. It could therefore be collected in a cistern and used as service water for toilet flushing. Since much of the water is retained and only about one third of the incident rainwater ends up in the cistern, it is normally not economical to install a second pipe network. Above and beyond the general relief of the drainage systems, rainwater retention by green roofs moderates extremes of temperature and improves the thermal insulation of the building during summer and winter.

Retention ponds

Retention ponds generally have beds sealed with pond membranes. Hence they differ from infiltration basins or swales in that they always have water in them. Designed as natural habitats, retention ponds have planted banks and provide living space for a rich variety of wildlife. The rainwater is mainly conducted by small watercourses from the roofs into the pond. The overflow from the pond during heavy rain is often taken to

Fig.64:
Rainwater drainage as a design feature

neighboring infiltration basins. Retention ponds can be a valuable feature in the design of public open spaces in residential developments.

Designing with rainwater

Elements of rainwater infiltration or retention works can also be used as design features to improve the utility value of open spaces and leisure parks. Instead of using underground drains, water is taken along open channels and streams, and can thus undoubtedly enliven the user's overall experience. › Fig. 64

›

\\Note:
Suggestions and advice on the use of water in open space planning can be found in: Axel Lohrer, *Basics Designing with Water*, Birkhäuser Verlag, Basel 2008.

USES OF WASTE WATER

In view of the fact that drinking water treatment is becoming more and more expensive and complex due to the increasing pollution, and yet high quality drinking water is only required in very few areas, it is incomprehensible that millions of cubic meters of rainwater and waste water enter drainage systems without being used. Recent years have seen an increasing number of concepts for substituting rainwater or gray water for drinking water.

Using rainwater

The use of rainwater saves drinking water and relieves the load on drainage systems and water treatment plants. It poses no risk to hygiene when used for WC flushing, garden watering, and washing machines, as long as it contains no heavy metals or other toxic substances. The quality of rainwater depends on the place where it falls and the character of the surfaces over which it flows. For example, the roof surfaces could be contaminated with street dirt or bird droppings and therefore rich in microbes. The water from highway or car park drains is unsuitable for further use because it may be polluted with gasolene or oil residues. In this case it is better to forgo using rainwater. It should always be possible to

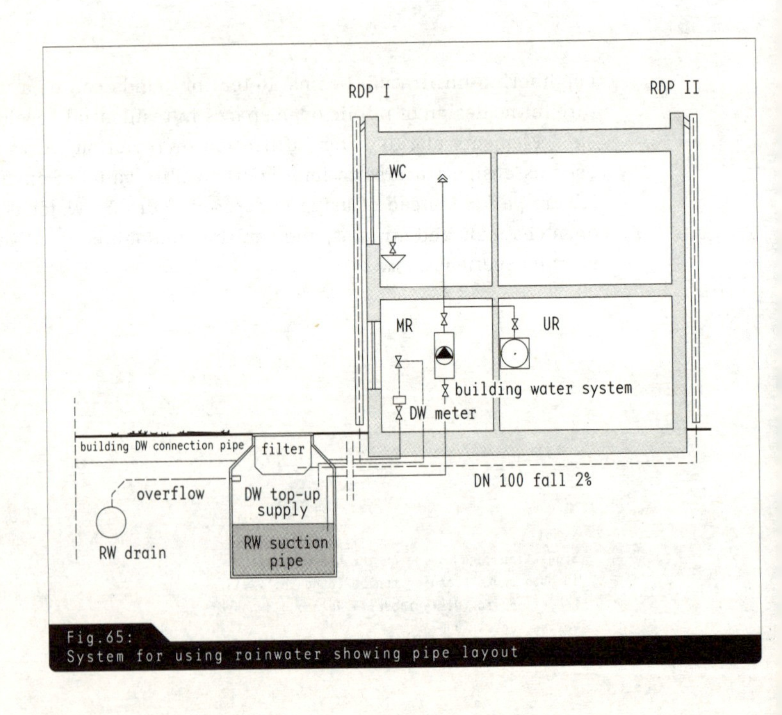

Fig.65:
System for using rainwater showing pipe layout

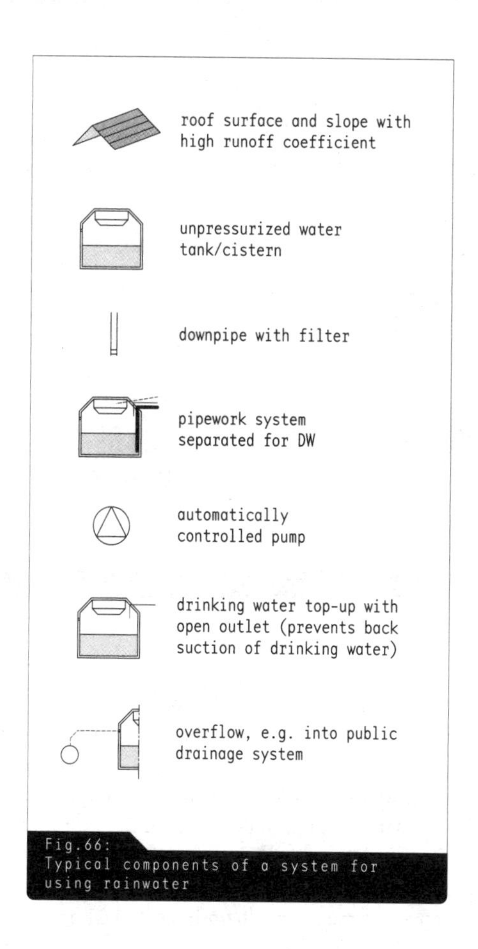

Fig.66:
Typical components of a system for using rainwater

design the components of any rainwater use system into a building with relatively little complication. › **Figs. 65 and 66**

Roof collection area

The size and characteristics of the roof surface used for collecting rainwater are critical to its collection and use. If the roof surface is smooth, a large quantity of rainwater will be able to drain off, but if it is constructed from a porous material, part of the water is absorbed and evaporated. All commonly available roof materials, such as clay tiles, concrete roof paving or slate, are suitable for collecting rainwater. Rainwater collected from metal roofs can cause graying of washing if a washing machine is connected to the rainwater usage system. There are no harmful effects if rainwater collected from metal roofs is only used for toilet flushing.

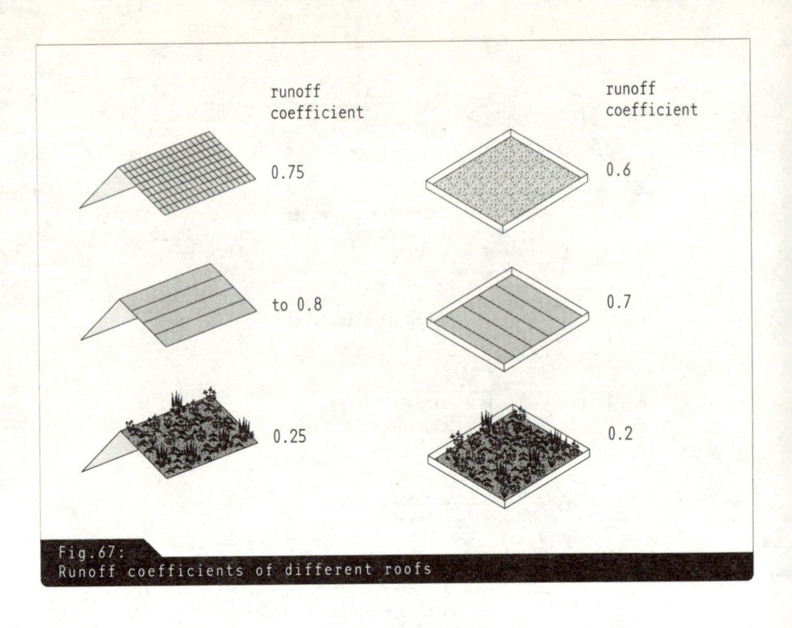

Fig.67:
Runoff coefficients of different roofs

The volume of rainwater delivered to the cistern depends on the intensity and frequency of rainfall events and the runoff coefficient of the roof. A runoff coefficient of 0.75 means that 75% of the rain falling on the roof flows through the downpipes and into the cistern. The runoff coefficient is between 0.0 and 1.0 and depends on the roof material. Smoother roofs have higher runoff coefficients. › Fig. 67

Collection tank cisterns

The collection tank is normally a cistern that accepts and stores water flowing off a roof through the downpipes and filters. Cisterns can vary in size and position. Buried cisterns are recommended if the building has no basement. Otherwise the tanks, manufactured from opaque plastic, are set up in a cool, dark basement room, so as to prevent the build up of bacteria and algae in the water. Cisterns are available in various sizes up to about 1000 l. For very high use, either several plastic tanks can be coupled together or a buried watertight concrete tank can be constructed to any size required.

Collection tank cisterns

The design of the storage capacity takes into account the rainwater influx and demand. Regional rainfall maps, which can be obtained from the meteorological office, are used for the calculation of the available rainwater. For example, in Germany the rainfall is between 600 and 800 mm per year depending on the region. The average length of a dry period is about 21 days. The calculation of the rainwater influx is based on the roof area and the runoff coefficient of the roof material.

›✎ Calculation of the annual rainwater influx in l/a:

Collection area (m^2) × runoff coefficient (w) × annual rainfall (mm/a)

Calculation of the service water demand in l/a:

Daily personal demand × no. of persons × 365 days

Calculation of the required storage capacity in l:

$$\frac{\text{Service water demand} \times 21 \text{ days}}{365 \text{ days}}$$

With a balanced relationship between rainwater yield and service water demand, a storage capacity of about 5% of the annual yield has proved adequate.

Drinking water supply and distribution

During dry periods, drinking water can be topped up from the public supply, through an open, frost-free outlet directly into the cistern, or inside a domestic water station in the building. A domestic water station contains a pump for the delivery of rainwater, controls, pressure regulator, and further safety devices. During topping up, the drinking water pipework must not be allowed to come into direct contact with rainwater, to protect the drinking water in the public supply. It must be possible to conduct away excess water that the cistern cannot accept, through an emergency overflow into the public drainage system.

Gray water recycling

In highly impervious inner city areas, where there is generally insufficient space for a reed bed waste water treatment system, a biological gray water treatment system is adequate for cleaning slightly soiled water. They are best installed in a building's basement and are usually an assembly of various system components, specifically chosen and arranged to suit the cleaning process. Which system is installed depends on the amount of space available, the number of users and the budget. With these systems, black water must be separated and led into the public drainage system.

✎

\\Tip:
For rainwater use, the roof area is calculated from a projected view, in this case the view of the roof from above.

Fig.68:
Biological immersion contactor system

Biological immersion contactor systems

A typical biological immersion contactor system consists of a sedimentation tank, a mechanical cleaning system in a holding tank, and a bucket wheel, which delivers the gray water continuously in a rotating biological immersion contactor. › Fig. 68 This is responsible for most of the cleaning action, and consists of polyethylene filter tubes, which represent the plant area in a microorganism carpet. The immersion contactor rotates at a speed of 0.5 revolutions per minute and is always half-immersed in the waste water tank, while the other half provides the microorganisms with oxygen by its contact with the air. The continuously growing biomass on the immersion contactor created by this process detaches itself from the rotating body in due course and sinks to the bottom of the tank. After the various purification stages have been completed, the cleansed waste water is no longer putrefactive and can be used as service water.

Membrane filter systems

Another option for gray water treatment is a membrane filter system, which first mechanically precleans the gray water through a ventilated mesh, using microorganisms and an oxygen feed to remove the organic constituents, and then cleans the water by passing it through several microfiltration membranes positioned closely one after the other. The membrane filter system is a closed system and can be installed simply and easily into a compact space in a basement room.

Gray water recycling is one of the most environmentally friendly waste water treatment processes. In addition to allowing more use to be made of rainwater, it also ensures that less drinking water has to be expensively cleaned and transported over long distances. Furthermore, the amount of waste water is reduced and the load on treatment works relieved.

If the treated water is allowed to infiltrate, it replenishes groundwater and contributes to the natural water cycle.

The waste water treatment and the environmental alternatives given here connect to the end of the water cycle since the cleansed waste water is returned to natural water bodies. Drinking water recovery and resupply to buildings takes its place at the beginning of the cycle.

IN CONCLUSION

This examination of the subject of service water shows that there is a wide range of variation in how we handle drinking and waste water, and furthermore that complex and careful planning is called for from architects to integrate a drinking and waste water pipe system, including the connected sanitary appliances, efficiently into their buildings.

However, there is still much more to be done if we are to achieve sustainable development in our use of water resources, as is presently demanded by energy-saving buildings. Comprehensive solutions are required to ensure the long-term stability of the natural water cycle and to avoid burdening it, even if temporary or long-term droughts only occur in rain-starved countries. Instead of developing more and more complex and expensive cleaning and treatment techniques for drinking water, we should be ensuring wherever possible that no pollutants do enter the groundwater. To achieve this will require complex measures that are outside the remit of the building designer.

In this respect, architects can exercise influence on how water is used when they advise clients and point out to them the range of possibilities within their buildings—not only concerning a beautifully designed bathroom, but also how water could used more sparingly and reduce the volume of waste water produced. Simple measures to reduce drinking water usage, such as water-saving faucets or more extensive arrangements for the use of rainwater, different ways of using waste water, and rainwater recycling offer environmentally friendly alternatives to conventional fresh water and waste water technology. Moreover, they promote the sustainable protection of our valuable drinking water resources. In the future, if on the grounds of costs and environmental protection alone, further developments in this area will place great emphasis on water and energy savings. Systems that use solar energy to heat drinking water support this principle. If the will is there, concepts for rainwater and gray water use can be implemented quickly and easily. Overall, these measures make a major contribution to the protection of watercourses and the stabilization of the water cycle, even if, individually, they may appear to have no great effect.

APPENDIX

LITERATURE

John Arundel: *Sewage and Industrial Effluent Treatment*, Blackwell Science, Oxford/Malden, MA, 2000

Tanja Brotrück: *Basics Roof Construction*, Birkhäuser Verlag, Basel 2007

Committee on Public Water Supply Distribution Systems, National Research Council of the National Academies: *Drinking Water Distribution Systems: Assessing and Reducing Risks*, National Academies Press, Washington, DC, 2006

Klaus Daniels: *Technology of Ecological Building*, Birkhäuser Verlag, Basel 1997

Herbert Dreiseitl, Dieter Grau (eds.): *New Waterscapes—Planning, Building and Designing with Water*, Birkhäuser Verlag, Basel 2005

Herbert Dreiseitl, Dieter Grau, Karl Ludwig (eds.): *Waterscapes—Planning, Building and Designing with Water*, Birkhäuser Verlag, Basel 2001

Gary Grant: *Green Roofs and Facades*, IHS BRE Press, Bracknell 2006

Institute of Plumbing (ed.): *Plumbing Engineering Services. Design Guide*, Institute of Plumbing, Hornchurch 2002

Margrit Kennedy, Declan Kennedy (eds.): *Designing Ecological Settlements: Ecological Planning and Building*, Cap. Water, Reimer Verlag, Berlin 1997

Heather Kinkade-Levario: *Design for Water: Rainwater Harvesting, Stormwater Catchment and Alternate Water Reuse*, New Society Publishers, Gabriola Island, BC, 2007

Axel Lohrer: *Basics Designing with Water*, Birkhäuser Verlag, Basel 2008

Frank R. Spellman: *Handbook of Water and Wastewater Treatment Plant Operations*, Lewis Publishers, Boca Raton, FL, 2003

Ruth F. Weiner, Robin A. Matthews: *Environmental Engineering*, 4th ed., Butterworth-Heinemann, Amsterdam/London 2003

Bridget Woods-Ballard et al.: *The SUDS Manual*, CIRIA, London 2007

TECHNICAL STANDARDS

EN 752	Drain and sewer systems outside buildings
EN 805	Water supply - Requirements for systems and components outside buildings
EN 806-2	Specifications for installations inside buildings conveying water for human consumption
EN 1717	Protection against pollution of potable water in water installations and general requirements of devices to prevent pollution by backflow
EN 12056	Gravity drainage systems inside buildings
EN 12255	Wastewater treatment plants, Part 5: Wastewater treatment plants. Lagooning process

PICTURE CREDITS

Photographs

All photographs by Doris Haas-Arndt

Drawings

Jenny Pottins
Simon Kassner
Helen Weber
Sebastian Bagsik
Indira Schädlich

THE AUTHOR

Doris Haas-Arndt, Doctor of Engineering, Visiting Professor of Technical Building Services and Building Ecology at the University of Siegen, Germany.

P9

导言

水管道系统是现代建筑中非常重要的一项建筑设施，是一个复杂的饮用水供给和废水处理系统。该系统是一个循环系统，与自然界中的水循环存在一定的相似之处：将收集的淡水通过管网系统供给建筑，并在需要的时候进行加热；洁净水通过管道导向浴室、厨房或者卫生间等房间的排水口，一旦洁净水从水龙头中排出即变成废水，然后沿排水管道排放到下水道中进行净化处理，最后排放到自然中再次变成自然水。建筑师们在进行建筑设计的时候必须整体考虑整个水循环，并对给排水系统进行仔细而周密的考虑，否则建筑中的洗手间将可能无法冲洗，洗衣机将可能无法操作，而淋浴头将可能没有水喷出。

接下来的几章内容将对整个水循环中的各个部分及其主要功能进行介绍。我们应该对给水系统如何工作、如何设计一栋建筑的给水系统以及设计过程中需要注意的问题等方面有一个清晰的了解。同时，本文对于废水的产生、排放原因，以及在给排水过程中可能遇到的主要问题和相应解决方法进行了相关说明。

P11

给水

地球表面大约有三分之二的面积被水覆盖，但其中仅仅只有0.3%是淡水，存在作为饮用水的可能。饮用水是非常高质量的淡水，适合人体的饮用和消耗。

P11

自然界水循环

自然界水循环——或者说水文循环——是一个不间断的连续过程：包括水分蒸发、降雨、雨水汇集到水体或者渗透入大地，并聚集成为地下水。在太阳辐射或者其他热影响的作用下，水分蒸发成为水蒸气，然后聚集形成云，最后再随着降水回到地球表面。雨水中的一部分被大地吸收，一部分蒸发到大气中，一部分通过毛细管作用被植物所吸收，另有一部分渗透到更深的底层中以维持地下水水位。〉见图1

地下水

地下水指的是存贮在某一不透水底层之上含水地层中，常年温度在8℃～10℃之间的降水。地下水一般来说是无菌的，并可以通过深井抽到地表上。人类的日常饮用水中差不多有四分之三是地下水，但在地下水进入公共给水系统之前需要经过若干净化和过滤工序处理。

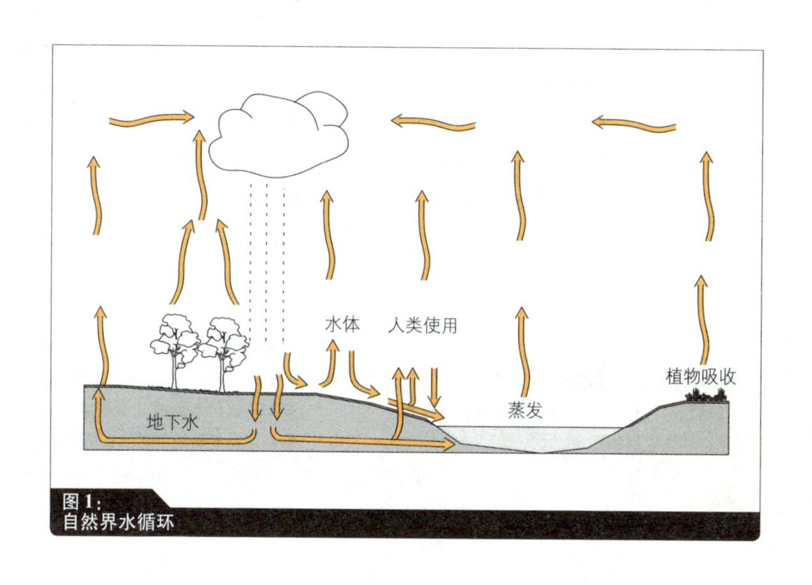

图1:
自然界水循环

大量的地下水抽取以及城市地面形成的土层大面积封闭对自然界水循环造成了巨大的影响。降落在不透水地面上的雨水无法自由地向底层渗入并汇集到地下水中，而是直接排放到地表水体或者排水沟中。由于城市建筑、森林的砍伐以及排水工程的增加，地下水位急剧下降。

另外，由于农业和工业需求所进行的地下水过度开采以及开采过程中所带来的污染，也对自然界水循环有害。农业肥料、杀虫剂、垃圾、公路排水系统以及工业排放物中的有害物质将随着酸雨落下并渗入到地下水中造成污染。为了消除这种污染，只能采用一些昂贵的净化和过滤工序。水污染的加重与用水量的增长之间的矛盾加剧，于是用水的成本也日益增加。

P12

饮用水标准

供人类使用的饮用水必须具备一定的标准，满足口感好、无臭、无色、无病菌以及无有害微生物等要求。每一个出水口都必须提供质量最好并具有一定水压的饮用水。对于欧盟国家，饮用水中所添加的用以消毒的化学物质和相关成分的用量必须满足欧盟委员会以及相关区域所制定的饮用水标准。根据现行的相关规范，需要经常对饮用水的质量以及其中所含物质进行检测。而规范中所规定的饮用水标准也是一直在改变的，目前所规定的最低污染标准只有通过大量的工序和消耗更多的成本才能达到。

水硬度

硬水指的是所溶钙、镁物质较多的水，而软水则指的是所溶钙、镁离子较少的水。硬度较高的水会导致管道中发生矿物质沉积；这些沉积的矿物质俗称“水垢”。采用硬水洗涤衣物需要使用更多的洗涤剂，而若洗碗机中使用的是硬水则可能在洗后的餐盘上残留一层较薄的石灰层。硬度的衡量单位是 mmol/L（毫摩尔每升），而硬度指标的高低主要由水的来源决定。〉见表 1 另一方面，当水中的碳酸氢钙含量小于 30mg/L 时也不能用于管道输送，此时无法在管道内壁形成保护层，

提示：

城市地面，比如沥青地面等，将对地层进行有效的密封，从而阻止水分的渗透，导致地下水无法得到有效的补充。

提示：

欧盟委员会 98/83 指令（欧盟饮用水指令）规定了饮用水的质量和要求，并强制其成员国在各自国家的法规中逐步实施该指令要求。

表 1：
水硬度分类

硬度指标	钙镁离子含量（mmol/L）	分类
1 度	<1.3	软水
2 度	1.3～2.5	中水
3 度	2.5～3.8	硬水
4 度	>3.8	极硬水

管道将受到酸性物质的侵袭而发生腐蚀。水的硬度对于人体健康的影响并不大。

pH 值

衡量水的“侵略性”的一种重要指标是水的 pH（拉丁语：potentia Hydrogenii）值。pH 值表示的是水中的氢离子含量，更准确地说是：水中氢离子浓度的负对数。根据这种测量方法，纯净水的 pH 值为 7，也就是说每升纯净水中的氢离子含量为 10^{-7}g。如果 pH 值小于 7，水将会具有酸性；如果 pH 值大于 7，水将具有碱性，同时将会有更多的石灰质发生沉积。

P13

饮用水需求

在 19 世纪时，每名德国人每天所需的饮用水和卫生用水的平均量为 30L。如今，由于卫生水平的提高（包括活水的使用、淋浴以及抽水坐便器的使用），该数据已经达到 130L。毫无疑问，该用水量已经非常大了，而且该用水量还是近些年来在卫浴间中安装了节水装置后有所降低的数据。然而，工业、商业以及农业用水量还在日益增加。从世界范围来看，农业灌溉用水占据了饮用水用量的大部分。

对于工业化国家，几乎所有的建筑都与公共饮用水供给系统相连。每年需要从自然水循环中摄取数十亿立方米的水作为饮用水使用，其中大部分都来自于地下水和地表水体，剩余部分则是通过河岸渗滤作用等方法得到的。“地表水体”指的是江河或者湖泊水，这些水一般都会被细菌或者固体污染物污染，在作为饮用水使用之前必须经过一系列繁琐的净化过程。

大都市区域的饮用水资源非常紧缺，其中部分饮用水需要从很远的地方输送。同时，由于城市中大部分区域的土层均被城市地面所封闭，绝大部分的雨水均被直接排放到了城市排水系统中。由于这些区域的饮用水供给非常困难，所以在这些区域降低饮用水需求势在必行。

家庭所需的饮用水可以用作不同的用途，其中真正被人体消耗的只是其中非常小的一部分。大约只有5L的饮用水用在供人饮用和烹饪中，剩下的都用在了其他地方。自来水厂中的储水设施可以对一天中用水量的波峰和波谷进行平衡和补偿。

家庭中每人每天所需的平均热水量大约在30～60L之间，居住者对于热水的需求量会因为各自的生活习惯在各天之间发生巨大的变化。一次沐浴需要消耗大约120～180L温度40℃的热水，而一次五分钟的淋浴则需要消耗大约40L温度37℃的热水。如果采用淋浴的方式代替沐浴，则可以节省一定的饮用水和能源。

P14

节约用水

如今，有许多方法和设备可以帮助我们节约用水：在淋浴头上安装流量控制阀，使用节水龙头和节水坐便器，使用具有节水功能的家用产品（比如洗衣机和洗碗机等）；另外，在每个房间安装流量表代替全屋仅安装一个总流量表的方式已被证明是一个有效的节水方法，因为居住者可以直接观察到自己的用水量，而他们仅仅需要对自己所用的水量付费。目前要求厕所中的水箱具有停止键，而且每次冲水所用的水量在4～6L之间。更加先进的设备，比如真空坐便器每次冲水所用的水量仅有1.2L，而无水厕所在使用后则不需要水进行冲洗。〉

见“建筑内部饮用水系统”一章中“卫生间”一节

表2：
饮用水标准用量

人类活动	用量（L/人/天）
饮用和烹饪	5
基本卫生用水	10
洗澡或淋浴	38
厨具清洗	8
打扫	8
衣物清洗	15
坐便器冲水	40
园艺用水	6
总计	**130**

对饮用水用量 〉见表2 进行进一步的深入分析后可以发现：真正需要使用的高质量饮用水只占总用水量的很小一部分——只有在个人卫生用水、厨具清洗、饮用和烹饪的时候需要使用纯净的饮用水。而雨水水质的水就足以供坐便器冲水、打扫或者园艺使用。将淋浴间和洗手池中的灰水进行净化处理之后，也可以用来冲洗坐便器。〉见“废水”一章中“废水利用”一节

若仅采用在卫生间安装现代节水龙头一项，就可以将人均每日用水量减少到100L左右。如果同时采用上文所提到的几项节水措施，人们有可能在不影响自我舒适程度的前提下将每日用水量减少到目前的一半。

P16

建筑内部饮用水系统

建筑内部的水循环一般来说是从与公共给水系统相连的冷水管道开始，除非该建筑小区有自己单独的供水设施（水井等）。在一些规模较大的城镇或者居住点，建筑内部管道和公共给水系统的连接一般都设置在道路两旁冻结深度以下距地面1.00～1.80m的土层内。每个小区都设有单独的饮用水供给管道，该管道与公共给水管道以直角形式相连，并向上升起，与建筑内部的接头、总阀门或者水表相连。〉见图2 对于住宅建筑，该水管的名义直径一般在25mm左右（DN 25）。

在一些欧洲国家，一般会在附近的建筑墙壁上采用有色标识表明饮用水接头，以便于判断和确定接头的位置。标识中的字母和数字将标明距饮用水接头的距离——从标识处为起点——以及接头所在的方位（位于左侧、右侧、前方或者后方）。标识中的另外一个字母缩写一般标识的是接头的类型，旁边的数字则表示的是管道的名义内径。

水温

为了防止饮用水中的微生物过度繁殖，要求提供给建筑的饮用水为冷水，水温要求在5℃～15℃。若想使用热水，则需要在建筑内部对饮用水进行加热。本书中的热水指的是水温在40℃～90℃的饮用水。从个人卫生用水的角度，适合的水温在40℃～45℃；而厨具清洗时则需要使用温度为55℃～85℃的热水以达到卫生的目的。

水压

为了将水供给到不同的位置，给水管道中的饮用水必须具有一定的压力。公共给水管道的水压一般在0.6～1MPa，在进入建筑内部管道系统之间，需要采用减压阀将水压降低到0.5MPa或者更低。〉见“建筑内部饮用水系统”一章中“饮用水系统组成”一节

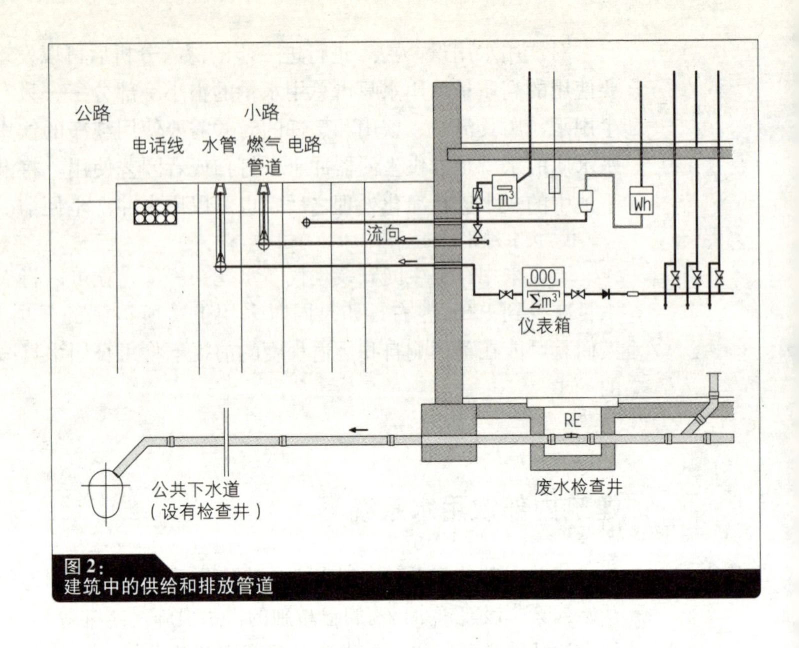

图 2：
建筑中的供给和排放管道

上述的水压值只能作为一个参考，具体将随着所在地区的不同而存在较大的差异。出水口的最低绝对水压不应该低于 0.05MPa，否则将不能有效地将水输送到所需的位置。给水管道和出水口之间的高度差等因素会导致管道中的水压损失，而出水口每升高 10m 则会导致大约 0.1MPa 的水压损失。

P17

饮用水系统组成

饮用水由建筑内部的水平和竖直管网输送到所需的位置，而这些管道一般可以埋藏在辅助管道、墙衬、楼板空隙以及墙槽内部。给水系统还包括记录用水量的水表、安全装置、截止阀以及出水口等。

提示：

标写在管道上的字母“DN”是“名义直径”的英文首字母缩写，其代表的是管道的内直径。其标写必须遵守各国的相关规范规定，具体数值由管道的尺寸决定。

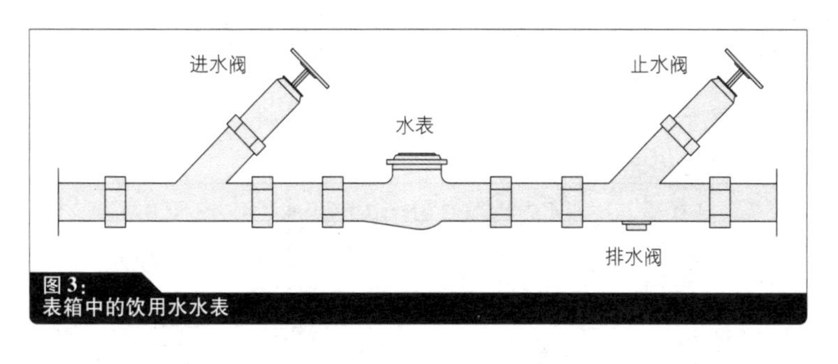

图 3：
表箱中的饮用水水表

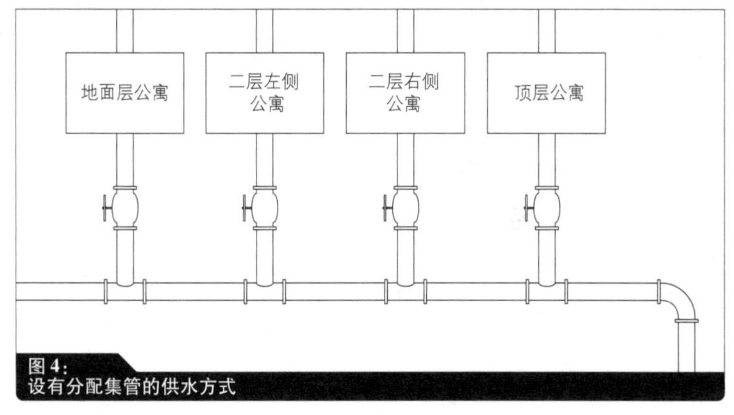

图 4：
设有分配集管的供水方式

进户管

与公共饮用水管道相连的进户管和水表等设施属于饮用水系统的组成部分，通常属于供水公司。进户管常常采用最短的路径与建筑相连，并且不允许在其上方建造建筑物，以便于进水管的铺设和避免对其维修带来困难。出于安全角度的考虑，进户管应该与建筑外墙或者基础成直角进入建筑内部，并在其外部设置保护套管。

水表

校准后的水表设置于公共饮用水系统供水龙头与建筑进水龙头之间，以便于其拆卸和安装。〉见图 3

从水表开始的部分属于建筑师的设计内容。水表需要放置于一个无结冻且便于达到的地方。比如，将水表放置于建筑通道边的仪表箱中，以方便对其进行读数。如果不能将水表安装在建筑内，则可以将其放置于建筑外部的仪表室中。有时可能会要求水表的安装能够让供水公司的工作人员在主人不在家的时候也可以方便地进行水表读数。

分配管

当供水系统需要设置多个立管的时候（比如，当需要对同一公寓的不同楼层进行分别供水的时候可能需要设置多个支管），需要设置分配管以对不同公寓进行供水。〉见图 4 除了通向不同公寓的支管之外，

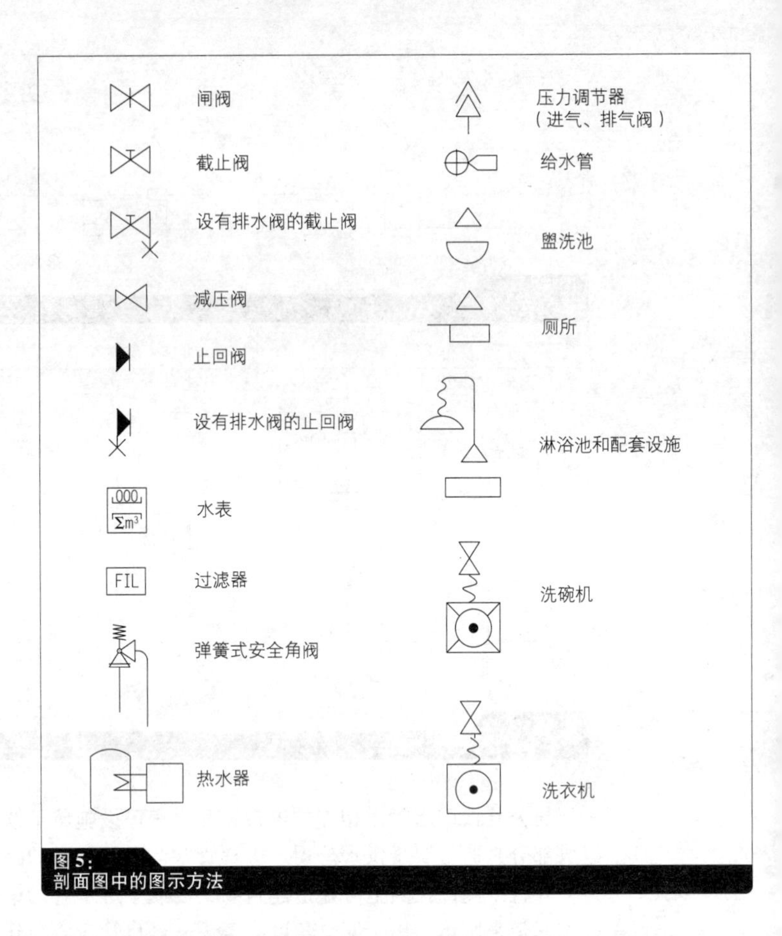

图 5：
剖面图中的图示方法

一般可能还要设置各自的热水管、出水口，而且在必要的时候可能还需要设置各自的消防用水。每一根与分配管相连的支管都必须进行清晰的标注，以便于分清不同支管所对应建筑物的不同部分。同时，每个支管都需要设置各自的止水阀，以便于在进行拆改的时候不会影响建筑其他部分的供水情况。

P19　布置图和图示

欧洲规范 EN 806 及其相关附录对设计图中用来表示供水系统组成和卫生设施的相关图示进行了明确的规定。采用这些图示可以标明建筑中需要安装的设施以及设施所处的位置、布置方式等内容。〉见图 5 不同国家对于图示的具体表示方法可能存在不同，但均要求通过平

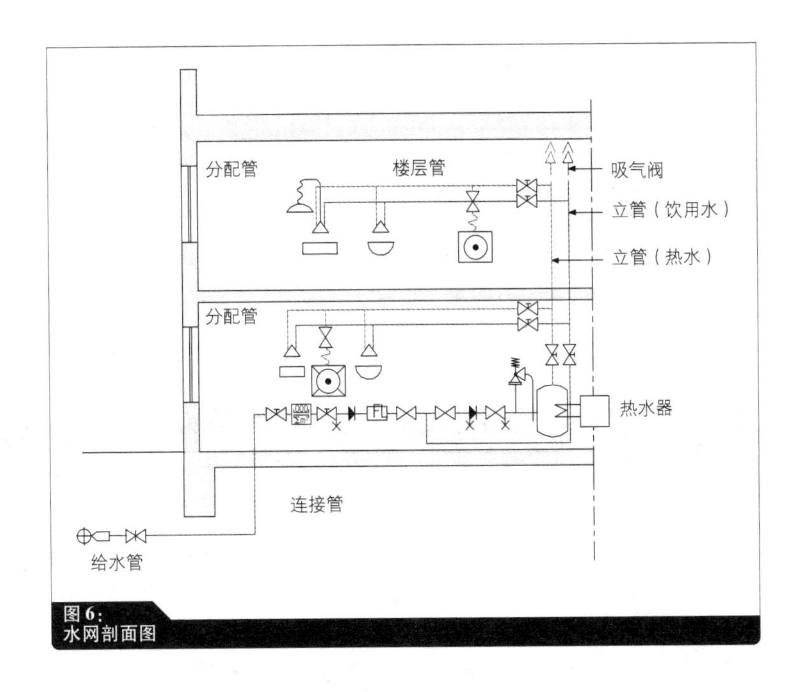

图 6：
水网剖面图

面图和剖面图的形式对整个饮用水系统和其管道进行准确描述。由于平面图中对有些卫生设施的表示视角是自上而下的，所以对于部分卫生设施，在平面图和剖面图中存在不同的表示方法。在设计图中设置图例索引，将更好地说明图示的含义。

剖面图表示方法

水网剖面图应该采用图示的方法，表达出尽可能多的关于饮用水系统的信息，其中的图示布置顺序应该与实际水网中的设备布置一致。剖面图将供水系统和出水口绘制在同一个平面中。〉见图 6

平面图表示方法

在进行平面图绘制的时候，非常重要的一点是在上、下水管线上标注箭头，以说明水流的方向。〉见图 7 另外，还需要对同层的水管进行标注，以区分是从本层开始的水管、结束于本层的水管或者是穿过本层的水管。〉见图 8

P20

管网组成部分说明

饮用水系统中的许多组成部分都具有与其他供给系统所不同的专用术语 〉见图 6，其中最常用的几个术语如下：

— 入户管，入户管指的是连接于供水系统和建筑主阀门之间的管道；

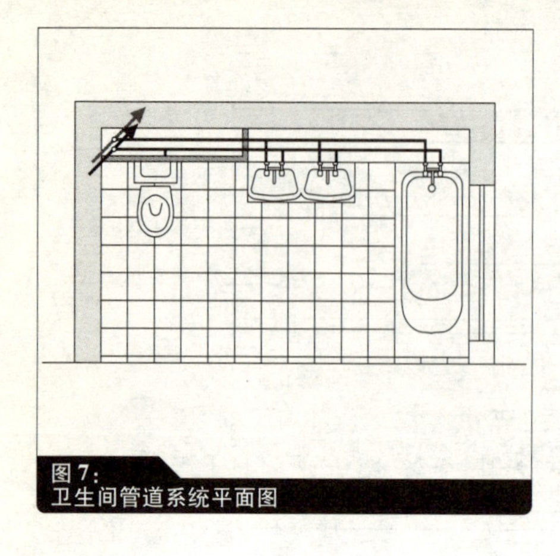
图 7：
卫生间管道系统平面图

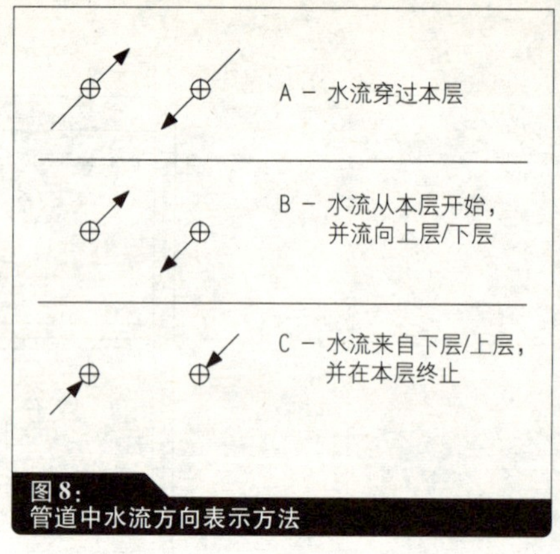

图 8：
管道中水流方向表示方法

— 立管，立管指的是竖直穿过建筑的管道，立管为各层的水平给水管供水；
— 水平给水管，水平给水管指的是从立管分支出来的水平管道；
— 循环管道，循环管道指的是为各出水口提供持续热水的管道，对于建筑而言，循环管道并非是必须设置的；
— 分配管道，分配管道指的是与各层水平给水管道相连，为各个出水口供水的竖向管道。

住宅中的立管一般采用 DN20 的管径，而各层的供水管道一般采用 DN15 的管径，也就是说它们的名义直径分别为 20mm 和 15mm。

循环管道

循环管道确保能够为各个出水口迅速提供热水，其优点是在出水口流出热水之前没有大量的热水流出，从而克服常见的瞬时热水器的弊端；而其缺点是需要消耗电能来驱动水泵，让热水在循环管道中持续循环。若在循环管道中设置一个定时自动开关，来控制水泵仅仅在需要热水的时候开启，则可以减少电能的消耗。

P21

管道线路

水平管道可以铺设在地下室顶棚或者楼层板内部。较大规模的建筑一般设有管道竖井，或者将管道设置在悬挂吊顶中。〉见图 9 位于地下室或者设备层中的竖向管道可以直接外露，与墙体相连固定。〉见图 10 对于位于管道竖井之外的长度较短的管道，则可以封闭于半高的假墙中。〉见“建筑内部饮用水系统”一章中“卫生间”一节

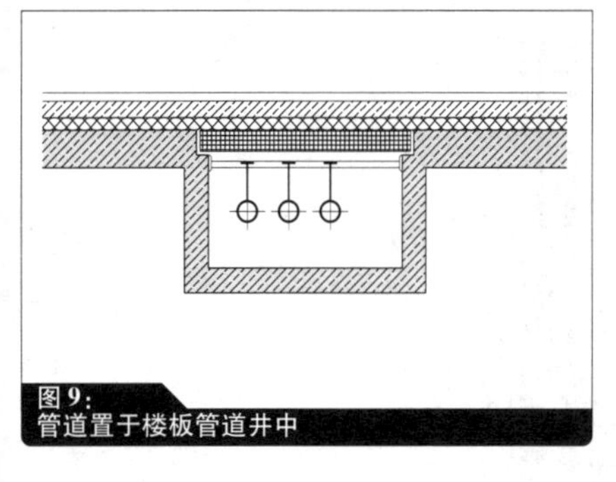

图 9：
管道置于楼板管道井中

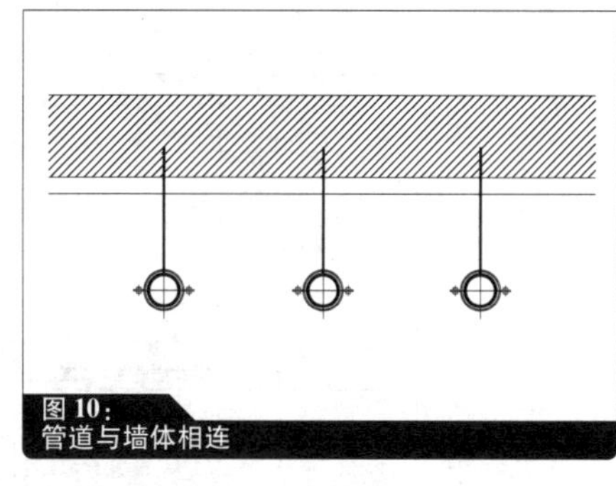

图 10：
管道与墙体相连

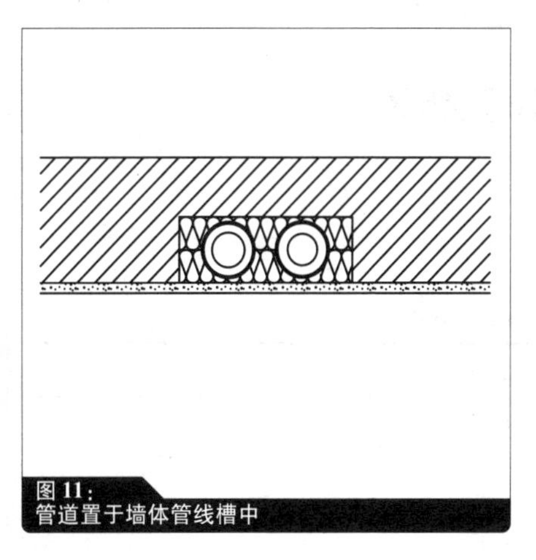

图 11：
管道置于墙体管线槽中

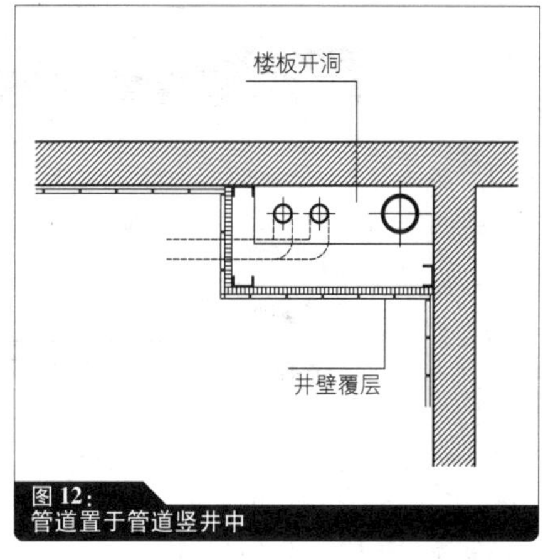

图 12：
管道置于管道竖井中

对于实心墙体建筑，如果墙体截面尺寸足够，而且从结构受力角度也允许的话，则可以将管道置于墙体管线槽中。〉见图 11 但由于存在施工过程复杂以及隔声效果差等缺点，这种方法逐步被设置于厨卫房间的管道竖井所取代。〉见图 12 与管道竖井相比，设置假墙的方式则只能覆盖墙槽中的管道，而且一般来说覆盖高度有限，而管道竖井则可以贯穿多个楼层。〉见图 13

P22

管道尺寸计算

管道的名义直径由与之相连的出水口数量确定。〉见表 3 与此同时，还需要考虑各出水口同时出水、管道所用材料、摩擦引起的水压损失

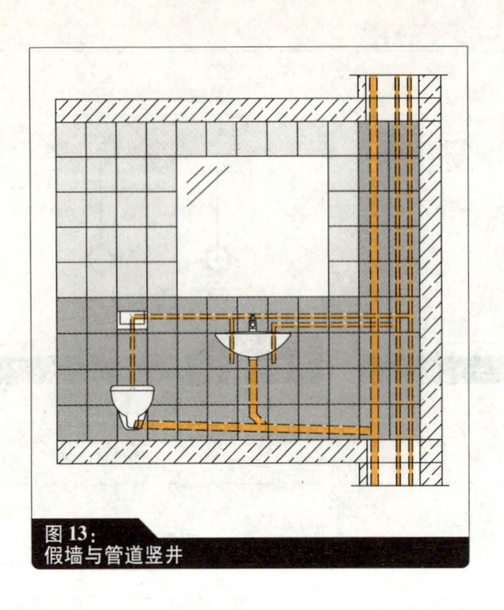

图 13：
假墙与管道竖井

表 3：
管道直径

管道类型	常用管道内径
入户管	DN25 ~ DN32
立管	DN20
1 ~ 5 个出水口	DN20
5 ~ 10 个出水口	DN25
10 ~ 20 个出水口	DN32
20 ~ 40 个出水口	DN40
楼层分配管	DN15
设有 1 个坐便器	DN10 ~ DN15
设有 1 ~ 2 个面盆	DN15
设有 1 个淋浴器	DN15
设有 1 个浴缸	DN20 ~ DN25
设有 1 个花园软管	DN20 ~ DN25

以及最小水压等因素。最小水压是为了保证管道系统中最远的出水口具有足够的水压来正常工作的值。管道的水压摩擦损失系数由一段管道中的水压损失除以其长度得到。

P24 管道材料

由于金属管道需要进行防腐保护，所以目前的室外地下饮用水管道一般采用聚乙烯塑料管道。另外一种比较先进的管道材料为复合材料，这种形式同时具有金属的强度优势和塑料的防腐性能。

室内饮用水管道可以采用铜管、镀锌管、不锈钢管以及塑料管等材料。塑料管由于韧性好、截面小的特点，可以以较小的半径进行弯折，甚至在楼板内部进行分布。塑料管通常设置成套管的形式——内层为柔软的交联聚乙烯饮用水管道，外层为韧性较好的高密度聚乙烯保护套管，在需要或者维修的时候可以将内层管道从外层保护管道中拔出并进行替换。由于其材料具有很强的变形性能，所以其抵抗水垢腐蚀的能力也很强。在使用金属管道或者进行管道系统部分改造的时候，需要特别注意的是务必将活性较弱的管道布置在活性较强的管道之后（相对于水流方向），比如将铜管布置在钢管后面，否则就可能发生腐蚀现象。

P24 安全设备

为了确保饮用水的品质，大多数的饮用水管道系统都设置了相关的安全设施。比如，防止非饮用水进入管道系统而降低饮用水的品质。安全设备包括防止水回流或者吸入污水，从而与饮用水发生混合导致饮用水质量降低。打个比方，当管道系统中某处生了渗漏导致水压降低，而同时淋浴头被放在盛水的浴缸中，就可能会使污水被吸到饮用水中。而安全设备可以防止水压降低所引起的污水回吸。

提示：

聚乙烯材料可以在埋管中采用；聚乙烯（PE-X）一般使用在内层管道中，外面再附加一层聚乙烯（HD）材料制成的保护套管。“HD”指的是高密度聚乙烯，“PE-X”指的是交联聚乙烯。

提示：

《欧洲规范》（EN1717）对室内饮用水的纯净度作了如下规定：“在管道系统中应采用必要的设备和措施，以防止回吸对饮用水造成污染”。

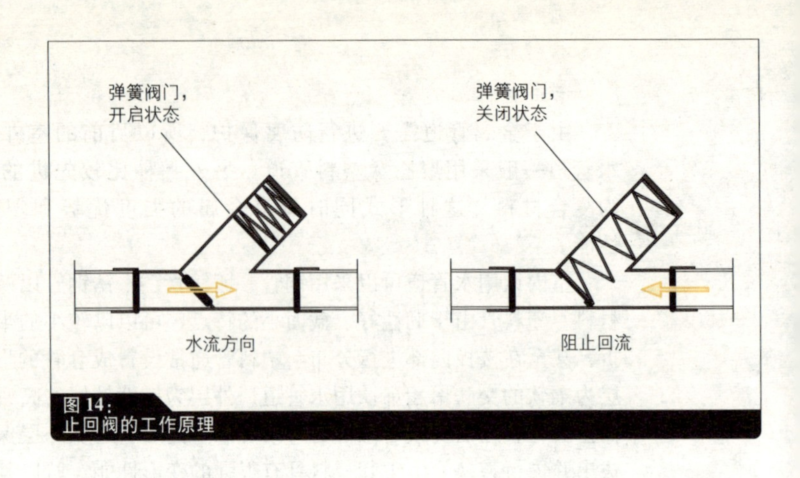

图 14：
止回阀的工作原理

止回阀

室内饮用水系统中最常见的一种安全设施即为置于入户水表之后的止回阀。止回阀为置于管道内部可以自动关闭的弹簧阀门，作用是防止管道中水的回流。只有在水流方向正确的时候止回阀才能开启。当水流停止的时候，止回阀自动关闭。如果水流方向相反，止回阀关闭并且压力增大。〉见图 14

吸气阀

吸气阀一般与止回阀同时安装，共同工作。其作用是在管道发生压力降低的时候吸入空气进行压力补偿，以防止压力降低造成回吸而对饮用水产生污染。吸气阀设置在每根冷水管或者热水管的最高处。通常情况下，吸气阀内部的阀门处于关闭状态。〉见图 15 当管道内部发生了压力降低的情况，吸气阀与止回阀同时工作。此时，吸气阀开启，空气进入管道内部，防止废水吸入管道内部。〉见图 16

当管道内部压力过高时，管道中的水可能会从吸气阀渗出，所以可以设置一段集水管与吸气阀相连，集水管与排水管相连，然后降水导入排水管流向下水道中。当吸气阀设置在淋浴头或者面盆上时，可以不必设置集水管，因为从中渗出的水分不会对管道系统造成损害。

减压阀

对于普通的出水口而言，从自来水公司直接提供的饮用水压力过大，所以可能需要在建筑内部设置一个减压阀来降低水压。水流压力作用在减压阀中一个可以活动的隔膜上，该隔膜可以按照不同的设置而处于开启或者关闭的状态。〉见图 17 减压阀的安装应该选择一个便于维修的位置。

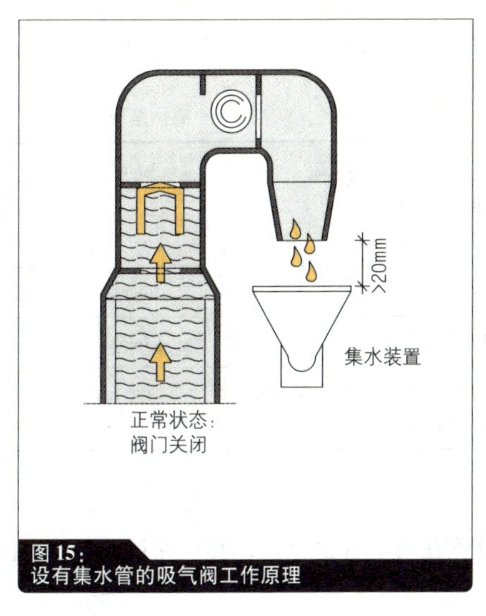

图 15：
设有集水管的吸气阀工作原理

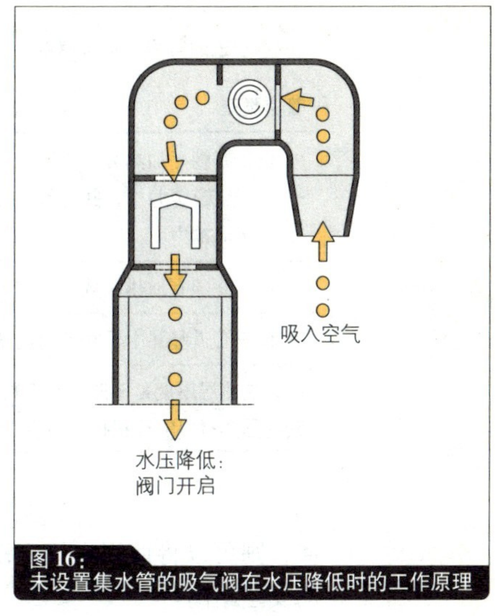

图 16：
未设置集水管的吸气阀在水压降低时的工作原理

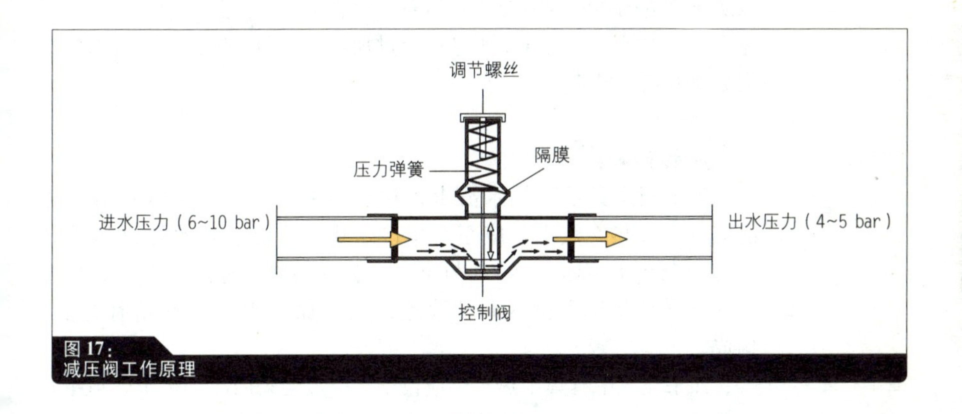

图 17：
减压阀工作原理

过滤器　　一个优质的过滤器应该确保清除饮用水系统的灰尘以及泥土颗粒。通常情况下，过滤器安装在入户水表和减压阀之间，以确保减压阀不被杂质所污染。由于安装过滤器不能确保饮用水中永远没有杂质污染，所以只有在能够确保定期维修清洗的前提下安装过滤器，才是一个值得并且有效的措施。

安全距离　　埋地式的饮用水管道以上 1m 范围内不得布置废水管道，以避免废水管道发生渗漏时对饮用水造成污染。如果无法保证 1m 的安全距

表4：
冷水管道的常用隔热层厚度

管道类型	隔热层厚度
— 位于开阔无加热区域 — 位于管道竖井中，并且附近无热水管道 — 位于墙槽中	4mm
— 位于开阔有加热区域	9mm
— 位于管道竖井中，并且附近设有热水管道	13mm

以上隔热层厚度是按照导热系数为0.035W/（m^2·K）的材料计算所得结果，对于其他隔热材料需要根据其自身性质进行重新计算。

离，则可以将饮用水管道设置在废水管道以上，并且保证20cm以上的安全距离。

管道隔热层

对于冷水管道，需要采取一定的隔热措施或者保持适当的距离，以防止附近的热水或者热水管道对冷水管道进行无意的加热，并且避免冷水管被污染。当饮用水管道穿过加热区域的时候，也需要对其采用必要的隔热措施。〉见表4

P27

热水系统

为了满足日常的热水需求，建筑内部的部分饮用水在分配之前进行了加热。热水供应系统由冷水供应装置、热水器（可能还含有一个热水存储容器）、与出水口相连的热水分配管以及循环管道（在某些情况下设有循环管道，以确保出水口能够随时提供）构成。

管道布置

如果建筑中有集中热水供应设施，一般会采用冷、热水管道平行布置的方式，而且二者的管道尺寸会大致相同。热水管道中水温一般

重点：

冷水管道务必设置在燃气管道的下方，以防止冷水管上的冷凝水腐蚀燃气管道而导致燃气泄漏，造成危险。

表 5：热水管道的常用隔热层厚度

管道名义直径	隔热层厚度
— ≤DN 20	20mm
— DN22 ~ DN25	30mm
— DN40 ~ DN100	同管道名义直径
— >DN100	100mm

对于长度不大于 8m 的管道，可以采用上述隔热层厚度的一半
— 在穿过墙体或者楼板的位置
— 在管道相交的位置

在 40℃ ~60℃之间。为了避免热量损失，热水管道应布置得尽可能短，而且在经过非加热区域的时候要采用必要的隔热措施。在进行管道布置设计的时候，需要考虑管道隔热层所需要的空间。

管道隔热层的厚度大致与管道的直径相同；对于长度小于 8m 的管道，其隔热层的厚度减少到一半即可满足要求。在管道穿过墙体或者楼板以及管道相交的位置，其隔热层厚度也可减少到一半。〉见表 5

单独、分组以及集中热水供应

建筑内部的热水可以采用集中供应、分散供应以及每个出水口单独供应等多种方式。如果某个热水器仅与一个出水口相连，那么这种供应方式称为单独供应；如果热水器与建筑内部部分出水口相连，那么这种供应方式称为分组供应；〉见图 18 和图 19 如果建筑内部的所有出水口均与单独一个中央热水器相连，那么这种供应方式称为集中供应。〉见图 20 在一个建筑内部可能同时存在单独、分散以及集中等多种热水供应方式。打个比方，在夏天的时候可以关闭建筑内部的中央热水器，而采用单独热水供应方式。

热水器

热水器大体上来说可以分为以下两种类型：一种是连续流热水器或者即热式热水器，这种热水器只有在使用的时候才对水进行加热；另一种是容积式热水器，这种热水器能够一直保持热水的温度而供人们随时使用。热水器可以根据其热源方式进行进一步的分类，热源种类包括固体燃料、石油、燃气、电能、地热能以及太阳能等。加热方式可以采用间接的热交换器和热交换介质进行，或者采用直接对水进行加热的方式。

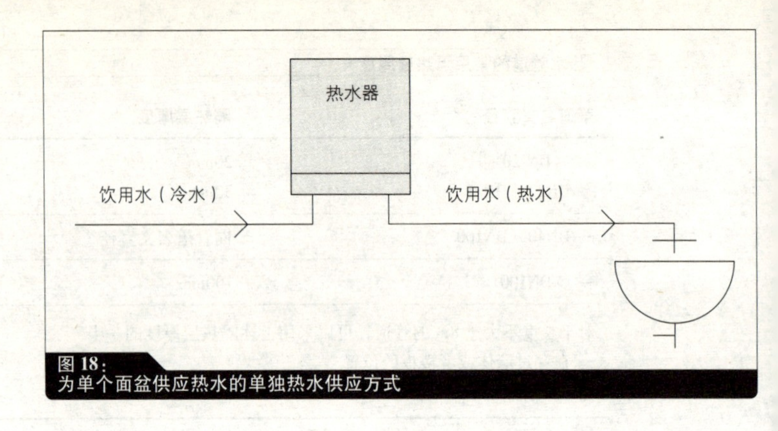

图 18：
为单个面盆供应热水的单独热水供应方式

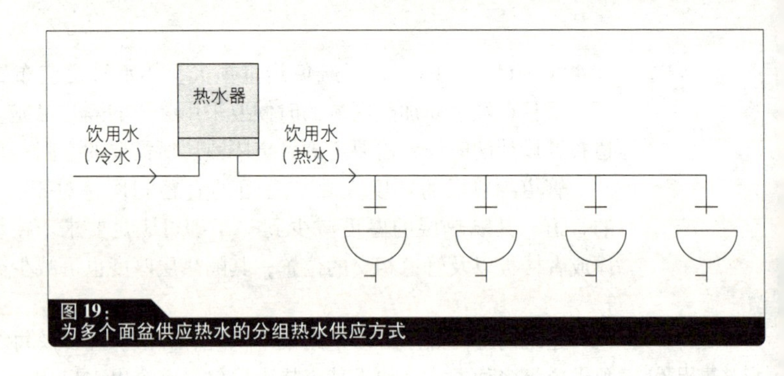

图 19：
为多个面盆供应热水的分组热水供应方式

如果有可能，设置在建筑内部的中央热水装置可以同时起到为建筑供暖和提供热水的功能。该装置由一个热水储水器和一个与之相连的加热装置组成，储水器用来存储热水和为加热回路提供热源，加热

提示：

热交换器的作用是将一种热交换介质中的热量传递到另一种交换介质中。比方说，水作为一种热交换介质，可以利用散热器将其中的热量传递到空气中。在热水供应系统中，热交换器置于热水储水器的内部。

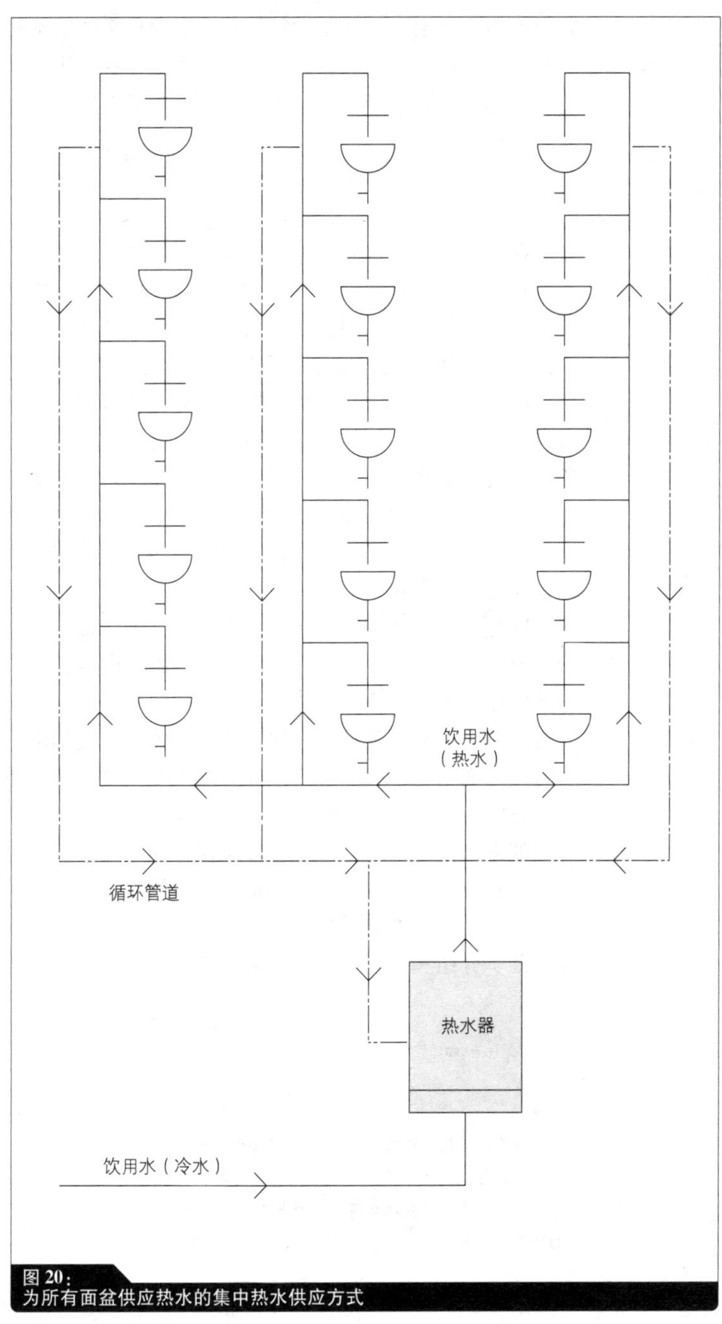

图 20：
为所有面盆供应热水的集中热水供应方式

装置则是对储水器中的水进行加热。采用水泵使热水和热能经过管道系统，流向各出水口和散热器。中央热水装置的优势在于可以连接太阳能加热器以及便于日后的维修和改进，另外还可以将加热装置中的热水输送到储水器。

P31 连续流加热系统

连续流加热系统同时也称为连续流加热器或者即热式加热器，这种加热器直接将水加热到60℃左右，〉见图21 其优点在于只需要对被使用的热水进行加热。相比于容积式热水器，循环加热系统不存在待机热损耗，而其中的热水通常不能被归为新鲜水。

连续流加热系统安装费用较高，但比较节省空间。由于其直接对水进行加热，故其加热效率较高。连续流加热系统的启动过程不可避免：也就是说在开大出水口之后需要一段延迟，流出热水之前将有一部分清水未加利用而随废水放到下水道中。从热水器到出水口之间的管道应该设计得尽可能短，以使流出热水之前的冷水尽可能少一些，但是无法完全避免。

即热式加热器可以采用电能或者燃气进行加热。电能加热一般采用交流电系统，而燃气加热一般需要设置烟囱或者烟道。电能加热器的能耗普遍较高，而且电能的生产成本较高，所以一般只有在水量较少或者的确没有其他替代能源的情况下才考虑使用电能进行加热。但从另一方面来说，虽然燃气属于矿物燃料，但在所有的矿物燃料中，燃气燃烧产生的二氧化碳最少。

如果采用的是组合式燃气加热器，则通过合理的设计可以同时提供热水和热能。热水器中的水量可以采用水压、温度或者电子设备进行控制。连续流加热系统能够同时为若干出水口同时提供热水，所以在单独、分组和集中热水供应系统中均可以采用。但是，连续流加热

提示：

矿物燃料包括煤炭、石油和天然气等。当它们燃烧的时候会产生二氧化碳（CO_2）并加重温室效应。同时，矿物燃料存在最终枯竭并不再作为能源的一天。

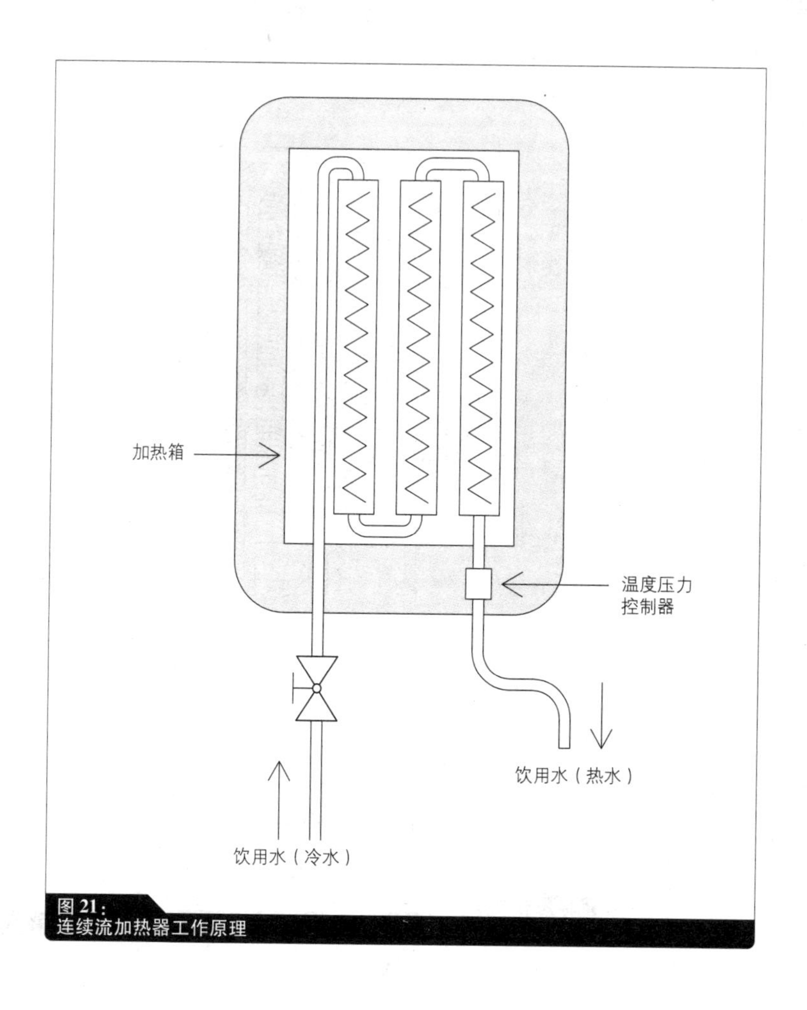

图 21：
连续流加热器工作原理

系统最大水流量存在限制。对于大量出水口可能同时用水的情况（比如宾馆或者体育场等），连续流加热系统的出水量则无法再满足要求。此时则可以采用中央容积式热水系统。

设有整体储水器的即热式加热器

这是即热式加热器的另一种变化形式，为即热式加热器配置一个储量为 15～100L 的热水储水器。如果需要使用更多的热水，比如浴缸沐浴的时候，剩下的热水则由即热式加热器进行即时加热。设有整体储水器的连续流加热器主要用途是为小型家庭单元提供热水和热能。

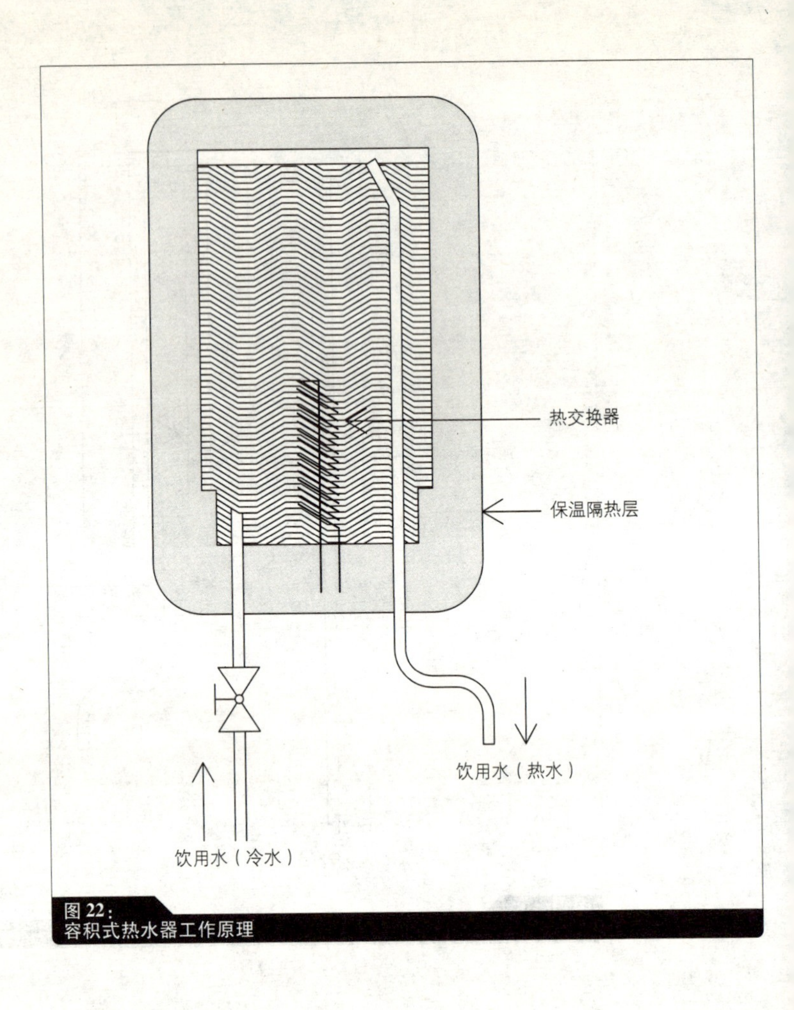

图 22：
容积式热水器工作原理

P33　容积式热水系统

容积式热水系统持续对储水器中的水进行加热，使水的温度保持在60℃左右。〉见图22 储水器中的水由热源直接进行加热或者由热载体进行间接加热，热载体可以是太阳能装置中的防冻液。〉见“热水系统”“太阳能热水”一节

热水储水器应该布置在建筑中央供暖设备间附近，并且需要进行隔热保护。内部封闭加压的热水储水器装置能够同时为多个出水口供应热水；开放式常压无隔热层的热水储水器一般用于为单独的出水口供应热水。

太阳能设备也可以用来改进容积式热水系统。由于容积式热水系统中的储水器和热水装置直接相连，当使用太阳能设备加热得到的温度不够或者当储水器中的水温降低到设定温度以下时，热水装置将对储水器中的热水进行再次加热。容积式热水系统的一个缺点就是如果储水器中的热水存储时间过长，经过反复加热之后的热水不再新鲜。

与即热式热水系统相比，中央容积式热水系统无法安装在出水口附近，因此需要设置较多的管道系统，从而其安装成本较即热式热水系统高。但是，从整体上来看，如果将热水系统和热能系统相连，在提供热水的同时能够为建筑供暖，那么整个建筑仅需要安装一个热水装置，中央容积式热水系统的总体成本将较低。

军团菌

对于军团菌而言，大型管道系统、设有大型储水器的饮用水加热系统以及隔热措施较差的饮用水管道为其提供了繁殖的温床。军团菌是一种棒状细菌，在冷水中的浓度很低，但在温水中会迅速繁殖。人类不会因为饮用而感染军团菌，但会因为吸入从含有军团菌的水中搅出的气雾而感染军团菌，比如在进行淋浴的时候。生病后的症状和肺部感染非常相似，包括发烧、肌肉疼痛、咳嗽以及呼吸短促等。在诊断的时候非常容易被误诊成为流感之类的疾病。如果生病后不能得到及时的诊断和采用适当的抗生素进行诊治，就可能会有致命的危险。

如果存储器中的热水温度降低到30℃～45℃之间并保持较长的时间，那么就存在军团菌污染的危险。避免军团菌污染危险的一个简单而有效的方法是对饮用水进行高温消毒，也就是每天或者每周将水加热到60℃以上，就可以杀灭其中的军团菌。另外一种方法是电子消毒法，也就是在水中加入消毒剂。

提示：

常压无隔热层容积式热水系统可以是热水装置或者终端热水器。这种装置大多安装在面盆或者水槽下方，适合需要迅速提供少量热水的场所，比如，办公室小厨房等。

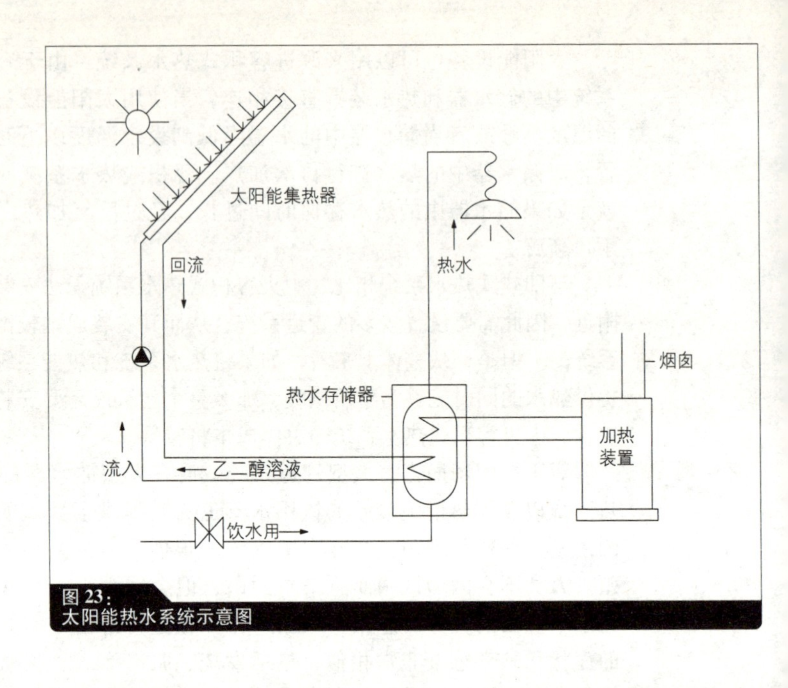

图 23：
太阳能热水系统示意图

P35　太阳能热水

太阳能集热器是热能利用的一种最环保的形式，不会产生任何形式的污染物排放。太阳热能装置主要用在热水供应方面，其供应量与天气和建筑的形式有关。如果太阳热能装置的面积增大，其提供的热能也可以用来为建筑供暖。

太阳能热水器由平板或真空管状集热器、液体循环装置和热水存储器组成。不同的太阳能集热器在制造成本和效率上存在差异，液体循环装置采用乙二醇溶液传递吸收的热能，而热水存储器则用来存储需要加热的水。热水存储器中设有一个热交换器，将吸收的太阳热能传递给存储器的水中。〉见图 23

在进行太阳能热水器设计的时候需要考虑所选集热器的类型，同时还要考虑热水器的用途是仅仅提供热水或者还需要为建筑进行供暖。

平板太阳能集热器

大体上来说，平板太阳能集热器主要组成部分是设有覆层的一层热能吸收体。其中覆层的材料比较特殊，能够允许几乎所有的太阳光透过其传到吸收体中。覆层由透明的无反射安全玻璃制成，该玻璃具

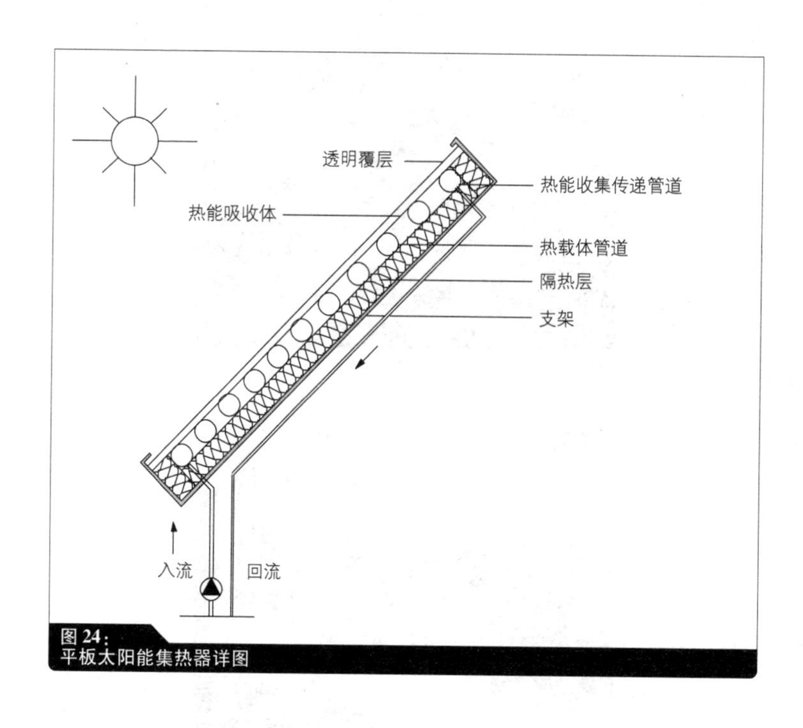

图 24：
平板太阳能集热器详图

有很高的太阳光穿透率，其背面和四周具有隔热功能。整个太阳能系统由其下方的支架支撑，玻璃层下方管道中的热传递介质将吸收的热能传递到热水存储器中。〉见图 24 ~ 图 26

✎ **真空管太阳能集热器**

为了提高真空管太阳能集热器的效率，其太阳能吸收层布置在真空玻璃管中。〉见图 27 真空管太阳能集热器由靠拢在一起的若干真空玻

✎

注释：

根据欧洲中部地区的气候情况，对于住宅性质的建筑，人均 1.2 ~1.5m² 的平板太阳能集热器就能满足日常的热水需求。而将该面积提高一倍之后，也就是人均 2.4 ~3.0m² 就可能同时满足建筑的供暖需求。

图 25：
安装于屋顶上的平板太阳能集热器

图 26：
安装于立面上的平板太阳能集热器

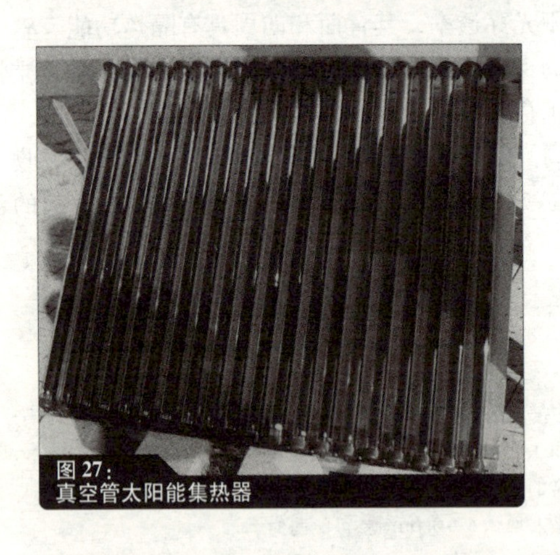
图 27：
真空管太阳能集热器

璃管组成，玻璃管由一个特殊的装置和装有乙二醇溶液的连接管相连。在安装的时候，可以通过旋转将集热器设置在一个特殊的角度，

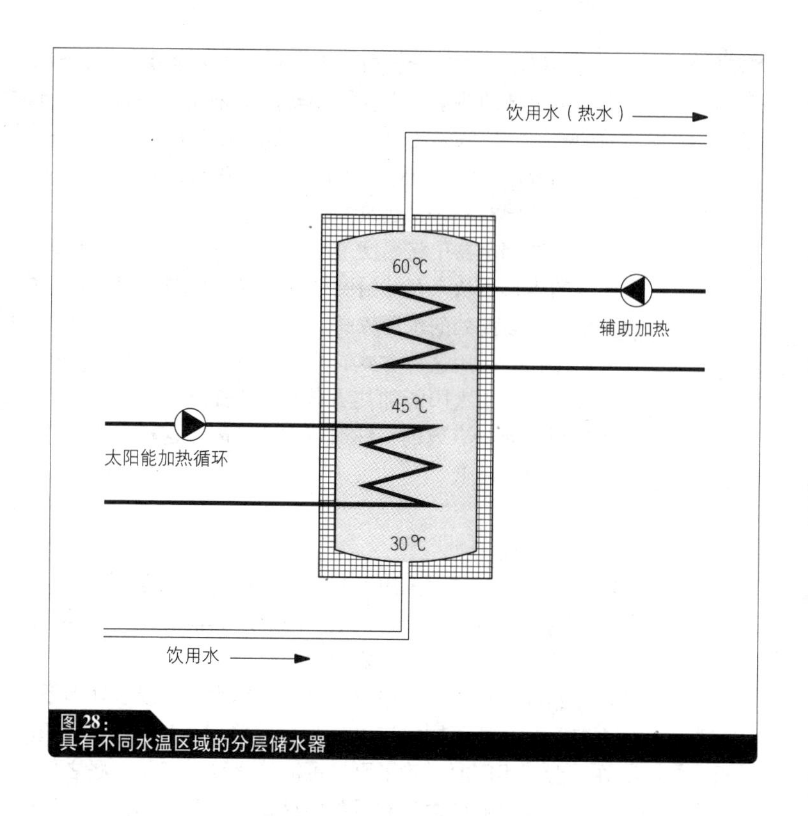

图 28：
具有不同水温区域的分层储水器

这样即使是在太阳能集热器安装角度不佳甚至是竖直安装时，也可以确保集热器有很好的效率。太阳能集热器的四周设有金属反光板，可以增加单位面积内吸收的太阳光。真空管太阳能集热器的效率相对较高，对于住宅建筑，人均 0.8 ~ 1.0m^2 的面积即可以满足日常的人均热水需求。

布置

对于坡屋顶，太阳能集热器一般布置在东南或者西南方向的屋面上；对于平屋顶，如果打算全年使用太阳能集热器供应热水，则一般呈竖向 30° ~45°进行布置。如果是欧洲中部地区，并且打算为建筑供暖，考虑到冬天的太阳高度角较低，则布置的角度应该达到竖向 60°。太阳能集热器，特别是真空管太阳能集热器也可以布置在阳台、建筑立面以及类似的地方。如果是因为建筑或者结构的原因，使得屋顶必须设计成竖直角度，或者无法设计成一个适当的角度，那么就需要为太阳能集热器专门设计一个布置区域。

覆盖程度

如果冬天太阳光辐射程度较低，太阳能集热器供应的热水可能无法满足全部的热水需求，那么就需要采用其他辅助的方法供应热水。一种建议的方法是将热水供应系统和建筑供暖系统统一起来，这种方法采用分层储水器方案将不同温度的热水存储在不同的水层中，在供应热水的同时也能为建筑供暖。〉见图 28

分层储水器中将温度最高的水存储在最上层，储水器的下层是饮用水的入口，从来存储温度最低的水。辅助的加热源布置在储水器的上层，所以此处的水温较高。当太阳辐射强度不足时，辅助的加热器开始工作。采用这种方式，大约 25% 的热能可以通过可再生资源得到补充。采用太阳能加热方式同时进行热水供应和建筑楼面（或墙体）供暖是最有效的一种加热方式，因为这种方式与散热器相比所需的给水温度较低。

P39

卫生间

热水供应可以采用上文所述的任何一种方式，然后采用与冷水管平行的管道输送到厨房、卫生间以及其他需要热水供应的房间，以供随时使用。卫生间的主要用途是个人卫生及护理，是一个建筑中使用频率最高的地方，其中设有冷水、热水以及废水处理系统。卫生间根据其设计形式不同，可以分为含湿区卫生间以及含湿屋卫生间两种。在布置卫生间时，除了要尽可能简化管道安装工序之外，还要尽可能减少管道竖井的数量和尽可能缩短给水管道和排水管道的长度。将管道进行成组安装不仅能够简化配管工作量，同时也可以减小管道传递到邻屋的噪声。

隔声

卫生间中噪声来源主要包括：坐便器蓄水池、排水管道中的废水排放、水龙头出水以及人在卫生间中的活动等。这些噪声可以通过墙体、顶棚和楼层板传递到其他的房间里。对于需要安静的房间或者卧

提示：

关于建筑供暖方面的进一步内容可以参考本套丛书的《空调设计》一书中“Tempering System”章节的内容，中国建筑工业出版社预计于2011 年 6 月出版。

室等，应尽可能避免布置在和洗澡间或者厕所相邻的位置。比如，若卧室的墙体中布设了相关设备，那么根本不可能将噪声完全隔离。一般来说，除非两个相邻房间之间采用了隔声连接，否则卫生间墙体的另一侧应该布置为厨房、浴室或者其他不怕噪声干扰的房间。

从声音传递角度来说，墙槽并不是一个很好的方案。虽然对墙槽中的管道进行隔声处理能够减小其噪声，但此时需要将墙槽深度增加，而与此同时常常会引起结构稳定方面的问题。

落地浴缸和便盆应该放置在橡胶隔垫或者浮式找平层之上，以防止产生的噪声传递到相邻的房间中。悬挂式卫生设备，比如便盆、面盆以及搁架等应该采用隔声或者塑料挂件进行连接，并布置在墙体密度较大的区域或者假墙上。

浴室中的水龙头和阀门具有两个不同的噪声等级：噪声较低的为 I 级；噪声较高的为 II 级。在需要进行隔声的地方，一般偏向于采用 I 级设备，不过有时可能会增加花费。

对于建筑师而言，卫生间的设计是一个比较困难的任务。在设计过程中不仅需要考虑房间布置、风格，同时还要考虑建筑隔声以及大量的管道安装等问题。这项工作需要细致入微的思考，否则随心的任意布置和设计可能会导致管道路线的不合理，并引起技术、功能以及经济方面等一系列的问题。

P40

卫生设备的布置

当建筑师在布置卫生间的时候，必须考虑卫生设备距离饮用水立管以及排水口的距离，并找到最简单直接的方法进行连接。由于饮用水管道的管径一般较小，故有时甚至可以布置在楼板之中；相比较而言，排水管所需要的管径相对较大，而且要求在建筑内部具有不小于 2% 的坡度，所以其布置会更加困难一些。对于不同的卫生设备，其排水管道的起点一般稍高于楼面，也就是说排水管起点和排水口之间的距离较短。

外露的饮用水管道和排水管道会产生较大的噪声。从另一个方面来看，不同形式的假墙和通向不同楼层的竖井可以起到密封管道和隔声的作用，用来取代造价相对较高的墙槽做法。

假墙设置

过去传统的做法是将卫生设备设置在实心墙体之上，然后采用砌设砌体将管道包围；而如今这种做法已基本被将管道布置在假墙后方的方式所取代，其优势是不仅不会破坏结构的稳定性，同时还具有较

图 29：
双面假墙的内部结构

好的隔声效果。在假墙内部一般设有一个金属支架和连接件，用于设置卫生设备，假墙内部的其他空间则用隔声材料进行填充，假墙表面可以采用石膏板进行覆盖。〉见图 29 假墙紧贴建筑墙体进行布置，其高度一般在 1～1.5m 左右，其厚度取决于管道的尺寸，一般在 20～25cm 左右。除非假墙与管道竖井相连，假墙一般仅仅密封本层的管道，对于楼层间的管道并不进行遮挡。假墙的上表面可以当做浴室架使用。

假墙的另外一种形式是采用整体模块组装体系。这种形式采用预制的密实聚酯泡沫混凝土砌块进行砌筑，将坐便器冲水器内置到假墙中并将其他卫生设备也设置到假墙之上。假墙的厚度一般在 15cm 左右，假墙可以直接设置在楼板结构层之上，或者在假墙和建筑墙体之间采用隔声防撞措施。假墙砌块之间的砖缝必须填实或者采用砂浆进行勾缝处理。

P42

卫生间设施

在确定卫生间的面积和选用相关设施时，首先需要考虑的是卫生间的使用人数和特殊需要。另外，浴室面积主要由其安装的设施以及相互之间的间距所确定。〉见图 30

如果在一个居住单元中有两个以上的居住者，那么采用浴室和卫生间分开设置的方式会为日常使用带来更多的便利。如果一个家庭中有两个以上的小孩，那么在已有浴缸和厕所的基础之上应该另外设置一个淋浴间和厕所。在确定卫生间布置和尺寸的时候，设计师应该考虑将卫生间的湿区尽可能布置在邻近的区域，以便于管道的安装而且避免设置较长的输水管道。

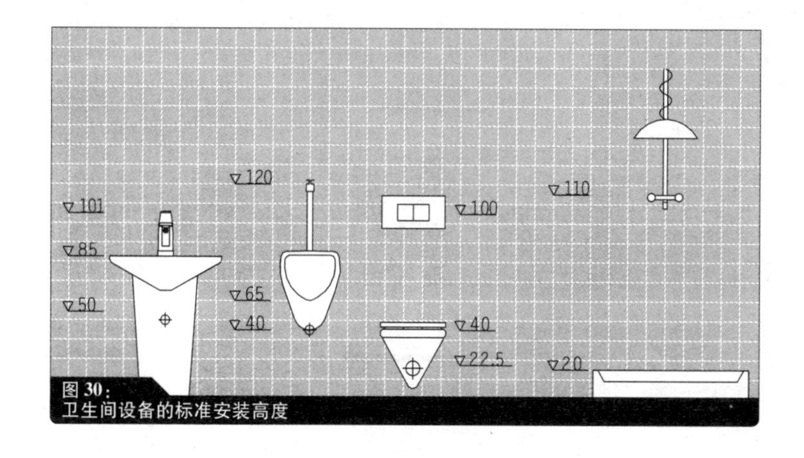

图 30：
卫生间设备的标准安装高度

在布置卫生间设施时，允许的最小相邻间距为 25cm，或者将其布置在不同的高度之上，以避免使用过程中的相互干扰。打个比方，面盆边缘和较低处的浴缸边缘之间在横向应该具有一定的间距。同样，在卫生设备前方应该留有一定的空地，以确保具有能够方便使用的空间。〉见图 31

面盆

洗手池和面盆的差别在于尺寸不同。卫生间的洗手池一般仅用于洗手，而面盆的尺寸较大。面盆的尺寸大小应该至少满足可以淹没手肘的要求，其材料大多采用的是陶瓷或者亚克力（有机玻璃），在个别时候也可能采用的是搪瓷或者不锈钢材料。洗手池和面盆的上边缘高度一般在地面以上 85 ~ 90cm 左右。与设置两个单独面盆相比，双面盆具有更好的经济性，但是要求双面盆的宽度不得小于 120cm，以确保两个人在同时使用的时候不会相互妨碍。〉见图 32

面盆的下方冷、热水管均设有角阀，以便于在维修的时候将水关闭。另外，也可以将上水管道、下水管道以及 U 形管 〉见“废水”一章“建筑内部排水系统”一节设置在浴室柜或者面盆下方的墙面之后进行隐藏。面盆也经常被设计成富有个性的浴室家具，以便于更好地利用浴室空间，并同时更加美观地隐藏管道系统。

淋浴设施

淋浴盆的常用材料包括搪瓷铸铁、搪瓷钢板以及亚克力等，并可以制作成矩形、方形、圆形、半圆形等多种不同的形状。〉见图 33 矩形淋浴盆的平面标准尺寸为 80cm × 80cm，高度在 15 ~ 30cm 之间；而尺寸越大，入口高度越低，则使用起来更加方便。如果要降低进入淋浴盆的台阶，那么需要将楼板结构层的凹槽深度增大，以便于排水管的连接，而排水管则可以在淋浴盆底部方便地进行安装。〉见图 34 上和中

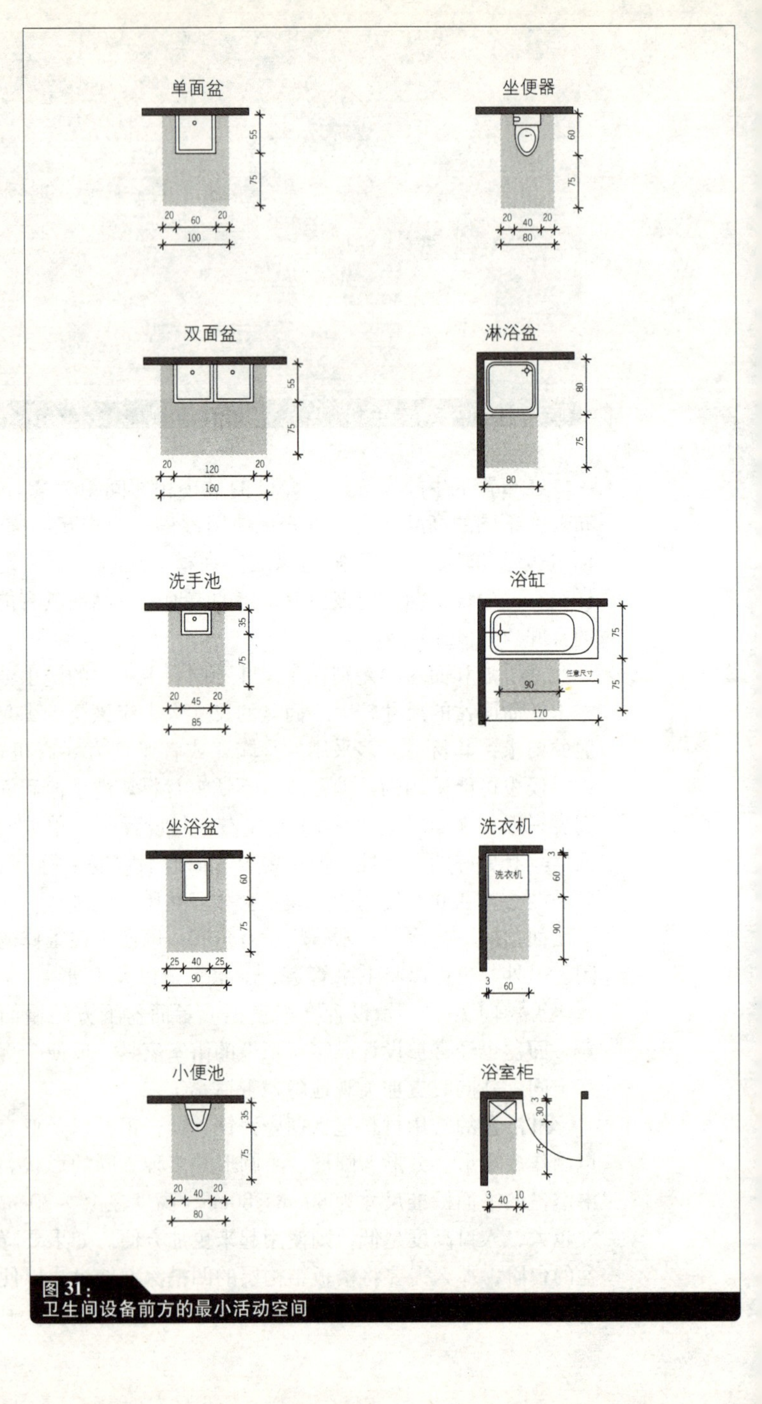

图 31：
卫生间设备前方的最小活动空间

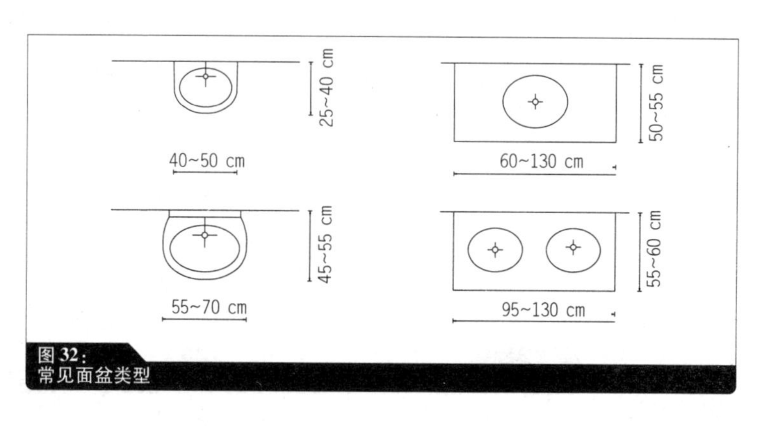

图 32：
常见面盆类型

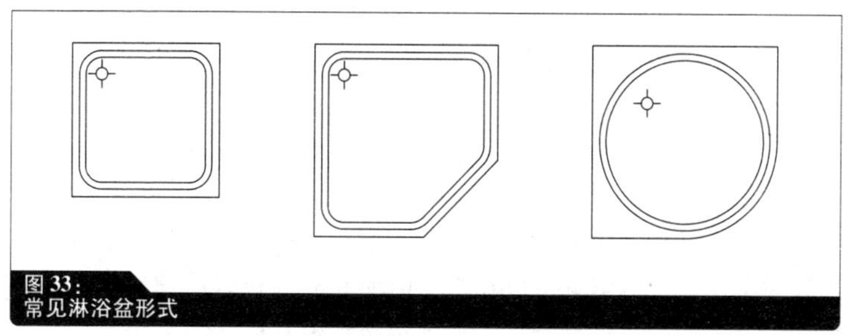
图 33：
常见淋浴盆形式

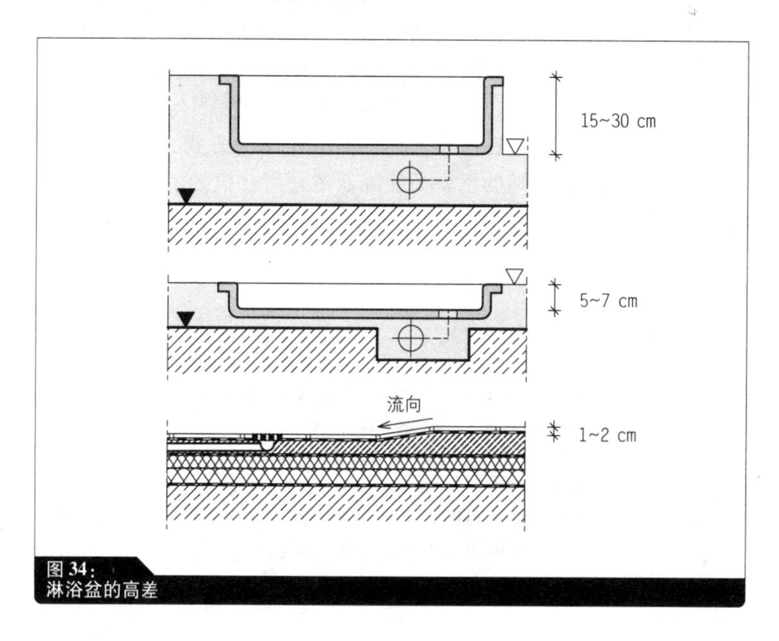

图 34：
淋浴盆的高差

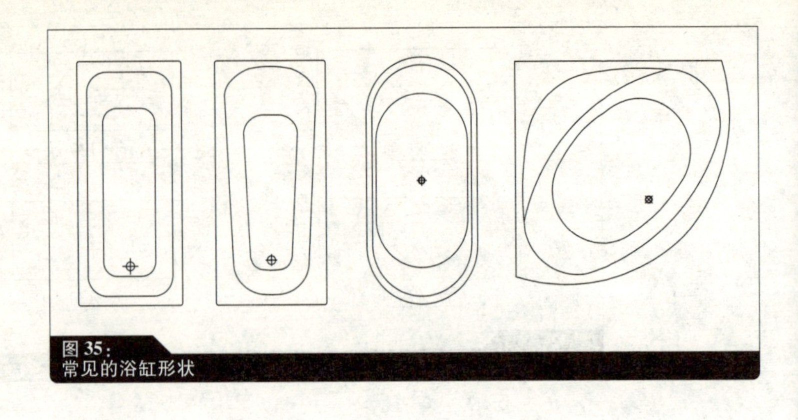

图 35：
常见的浴缸形状

有的浴室将淋浴盆和地板设计成为一体，淋浴盆的高度和浴室地板高度相同。〉见图 34 下图这种设计不仅具有美观方面和卫生方面的优势，更是一种无障碍设计。但是这种浴室的排水口位于楼板中，故要求确保相应的防水和封边质量，同时为了确保排水坡度，楼板的厚度相应较大。

浴缸

浴缸可以在浴室中独立靠墙放置，周边用平板进行覆盖，或者放置在预先设置的泡沫塑料模具中。制作浴缸的材料可以是铸铁搪瓷、搪瓷钢板或者亚克力等，其长度一般在 170 ~ 200cm 之间，宽度在 75 ~ 80cm 之间，高度在 50 ~ 65cm 之间。如果需要降低进入浴缸的高度，那么可以将浴缸嵌入楼板中，并要确保排水管的有效连接。通过在楼板上设置浴缸底座的方式，可以使浴缸成为可冲洗浴缸，否则的话就只能降低浴缸所在位置的楼板高度。

提示：

浴室的防水和封边可以选用防水膜和密封带或者其他防水涂料。防水涂料一般是采用薄层工艺涂刷在基底板之上。涂料应该涂刷到楼面 15cm 以上的高度，对于淋浴头可能溅到的区域，即使淋浴头在浴缸之上，也需要将防水涂料涂刷到更高的高度。在淋浴头所在的墙面位置，防水涂料必须涂刷到淋浴头以上不小于 20cm 的高度。

图 36：
架设在墙体上的便盆和放置于楼板上的便盆

图 37：
浅冲式和深冲式坐便器

图 38：
架设在墙体内部的水箱和放置于楼板上并设有压力室的坐便器

通常在浴缸和四周的覆层之间的空隙中设置隔声材料，浴缸和墙体之间的缝隙一般采用柔性封条进行防水处理。在浴缸长度方向的一段应该留有 90cm×75cm 的空间，以便于进入浴缸。

便盆

便盆可以架设在墙体中或者放置在楼板上。〉见图 36 将便盆架设在墙体之上可以为地板的清洁带来便利，这种做法通常是采用暗藏在假墙中的支架架设便盆，故可以对便盆的高度进行调节。便盆的形状一般根据冲水的方式决定，在老式的建筑中主要安装的是浅冲式便盆，而如今使用越来越多的是噪声较小的深冲式便盆。〉见图 37

冲洗水箱

冲洗水箱的形式包括设置在墙体内部的水箱、外露式压力室水箱、镶嵌式水箱、内置水箱以及联体水箱等。〉见图 38 水箱可以设置在便盆以上不同的高度范围内。老式的建筑中常用的是高空水箱，这种水箱在使用时会产生较大的噪声。故在新建的建筑中采用的大多是低置或者是镶嵌于墙体内部的水箱，可以大大降低噪声。压力水箱可以利用饮用水的管道压力进行工作，因此其形状并不需要设计成常见的箱型。压力水箱中的自闭式阀门能够确保在需要的时候随时可以进行冲水，也就是说只要按住冲洗阀门就一直能够冲水。而传统的水箱则是在每一次冲水之后自动对水箱进行再次注水。

厕所节水系统

日常生活所使用的饮用水中有超过三分之一是用在厕所冲洗上，也就是说平均每人每天要使用 35~45L 的水来冲洗厕所。减少厕所冲洗的消耗将非常有效地节约饮用水用量。老式的冲洗水箱每次冲洗所使用的水量在 9~14L 之间，而如今新型的冲洗系统每次冲水仅仅消耗大约 6L 水。通过调整水箱中的注水高度可以设定冲水量，也可以采用二次按压冲洗按钮（节水按钮）的方法来中断冲水。如果要将每次冲水的用量降低到 3L 左右，则需要采用特殊的便盆，否则使用普通的便盆会产生较大的噪声。

真空坐便器

真空坐便器在现代高速火车以及轮船上已有一段使用历史，而其每次冲水仅需要 1.2L 水。在建筑中则是采用水泵将坐便器中的物体吸出，再输送到通风的废水容器中。然后，采用另外一个水泵将废水容器中的物体输送到公共下水道中。在减少饮用水用量的同时，也大大降低了废水处理的费用。由于真空坐便器中的管道系统直径较小，所以使其在安装过程中不会遇到麻烦。但是，真空坐便器在冲水的过程中所产生的噪声要比其他类型的坐便器大。

无水坐便器

无水坐便器在冲洗的过程中不需要用水，所以也不会产生废水。在考虑生态因素或者当厕所无法与下水道相连的时候，可以使用无水坐便器。无水坐便器的堆肥箱设有两个通道，一个与坐便器相连，另

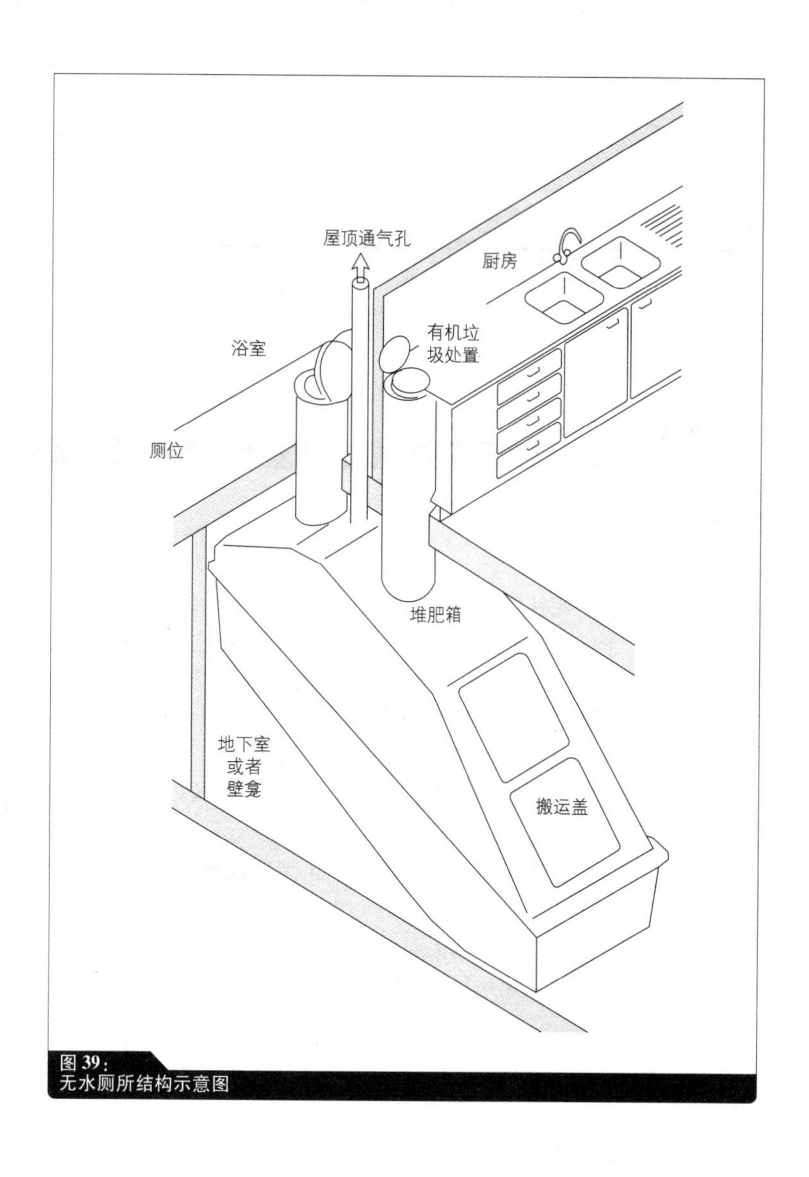

图 39：
无水厕所结构示意图

一个用来搜集厨房中的有机废物。〉见图 39 堆肥箱一直处于低气压状态，确保不会有任何气味传到室内空气中。堆肥箱中的空气流动可以促进其中废物历经数月的腐烂分解，而腐烂分解过程中所产生的养分则可以用来改进土质或者用作花园肥料。

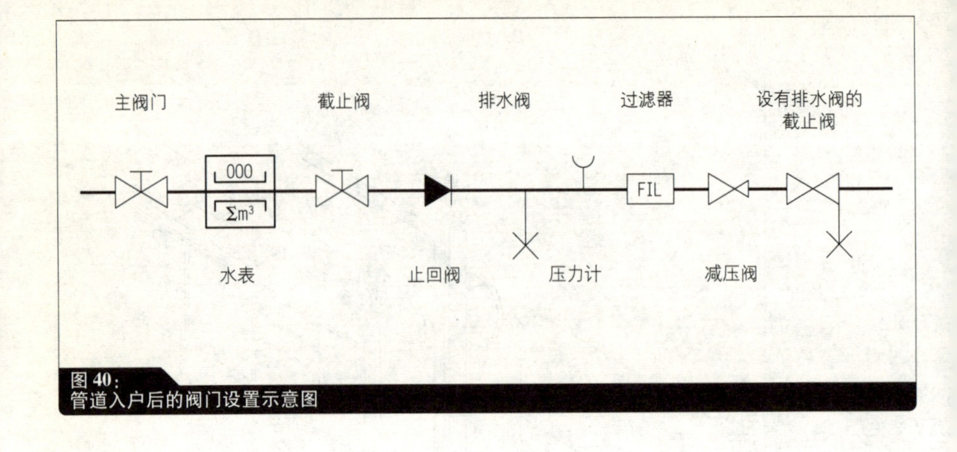

图 40：
管道入户后的阀门设置示意图

P49

配件

此处所说的“配件”包括饮用水系统中所有的阀门和卫生设施上的相关配件，比如水龙头、闸阀、截止阀，以及面盆和淋浴管的相关配件等。阀门用来控制饮用水在整个管道中的流动情况，但在关闭水流的方式上存在不同。截止阀将饮用水系统分割成不同的几个部分，以便将不同的部分进行隔离和更换。因此，截止阀适合设置在水表、过滤器以及减压阀前后，〉见图 40 还可以设置在每个立管和各层供应管的最低位置。这样的话，在进行部分管道更换的时候就不会影响这个饮用水系统。另外，截止阀还设置在厕所冲洗水箱中以及面盆下方。

墙面和表面安装水龙头

卫生间的相关配件可以安装在墙面或者平台表面上。〉见图 41 墙面安装水龙头一般主要用在浴缸和淋浴设备上，可采用一个短小的连接

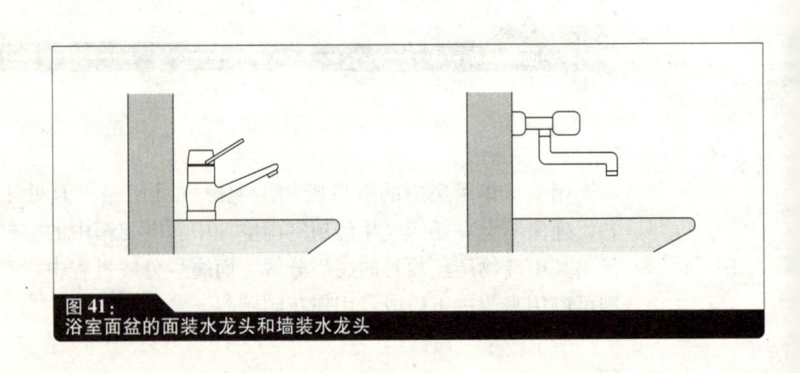

图 41：
浴室面盆的面装水龙头和墙装水龙头

装置将水龙头直接在墙面与水管相连，并同时起到密闭的作用。表面安装水龙头一般直接设置在面盆或者水池的上表面，并通过角阀与饮用水管道相连。水龙头的类型需要根据其用途来选用，比如厨房水龙头的管口一般比浴室水龙头长。

混水器

传统的混水器一般含有两个相邻的可以旋转的把手，其左、右把手分别用来控制热水、冷水水流，通过调节不同的水流来达到合适的温度。而实用性更强的是单把手混水器，通过上下旋转把手来控制水温。

感应式面盆水龙头

从卫生的角度考虑，感应式面盆水龙头一般安装在公共卫生间中。通过手的活动，就可以激发红外线传感器打开水龙头。对于一些电子控制水龙头，只需要将手放到水龙头附近就可以激发感应打开水龙头。水龙头中还安装了控制阀，用来确保每次激发感应导致的水流量相同。感应式面盆水龙头需要有电能的提供才能正常工作，可以通过安装电池或者设置额外电源的方式来确保其正常工作。

恒温水龙头

恒温水龙头通过温度调节钮进行温度预设，以保证水龙头即使在水流变化的情况下的出水温度仍恒定不变。恒温水龙头通过调整冷热水的比例来确保出水达到预设的温度值。〉见图 42

p51

无障碍卫生间

无障碍卫生间需要满足一些特殊的条件。首先，无障碍卫生间要确保使用者在使用其中所有设备时均不需要其他人的帮助。为了满足这个要求，需要确保在面盆、坐便器和浴缸前方具有不小于 120cm × 120cm 的无障碍自由空间；对于轮椅使用者，该区域不得小于 150cm × 150cm。为了让轮椅使用者更加方便，可以将淋浴盆设置成与楼面同高的无台阶淋浴盆，同时面盆的设置应该确保轮椅能够驶

提示：

在设置所有水龙头和卫生设备时应该考虑室内面砖的铺设网格，以保证美观的效果。水龙头一般应该避免设置在面砖接缝、角部或者中心位置。

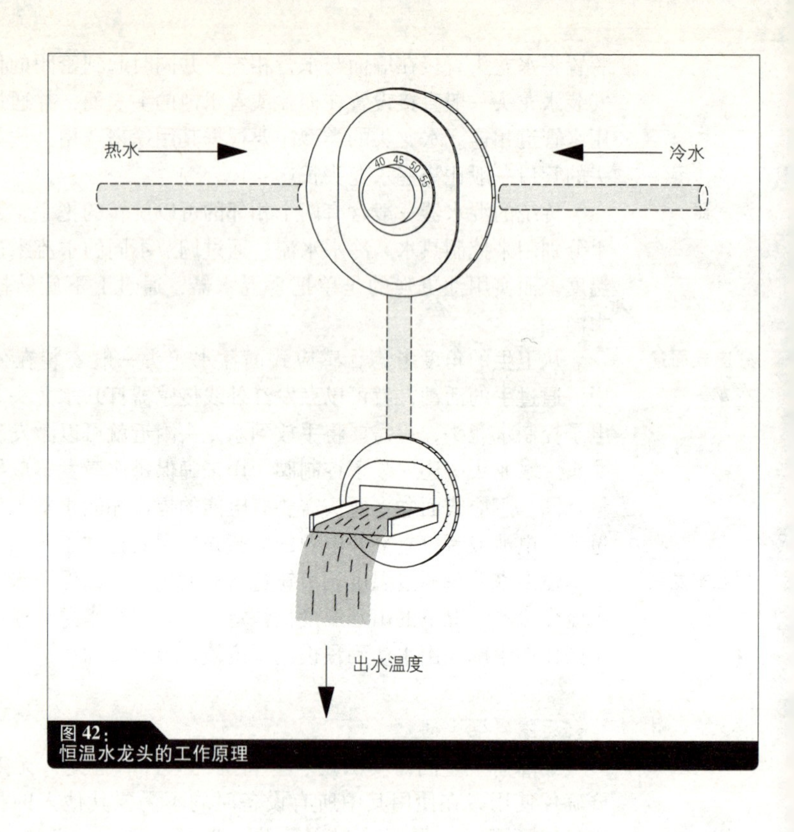

图 42：
恒温水龙头的工作原理

到其下方，另外应该在所有卫生设施的附近设置扶手。卫生间门的宽度应该不低于 80～90cm，并且朝外打开，避免在进入卫生间的时候影响室内的自由空间。〉见图 43

P53

废水

一旦饮用水经过建筑中的饮用水管道，从出水口处的水龙头排到面盆、淋浴间或者浴缸中，即使是干净的而且完全未经使用，也立刻变成了废水。一旦水进入了排水管道，“饮用水”的概念立即不再适用。

“废水”的概念不仅涵盖生活、商业以及工业废水，也包括相对洁净的降水（雨水）等。废水中一般含有固体颗粒、细菌以及化学物质等污染物，故在进入自然界水体之前需要进行相关处理。废水处

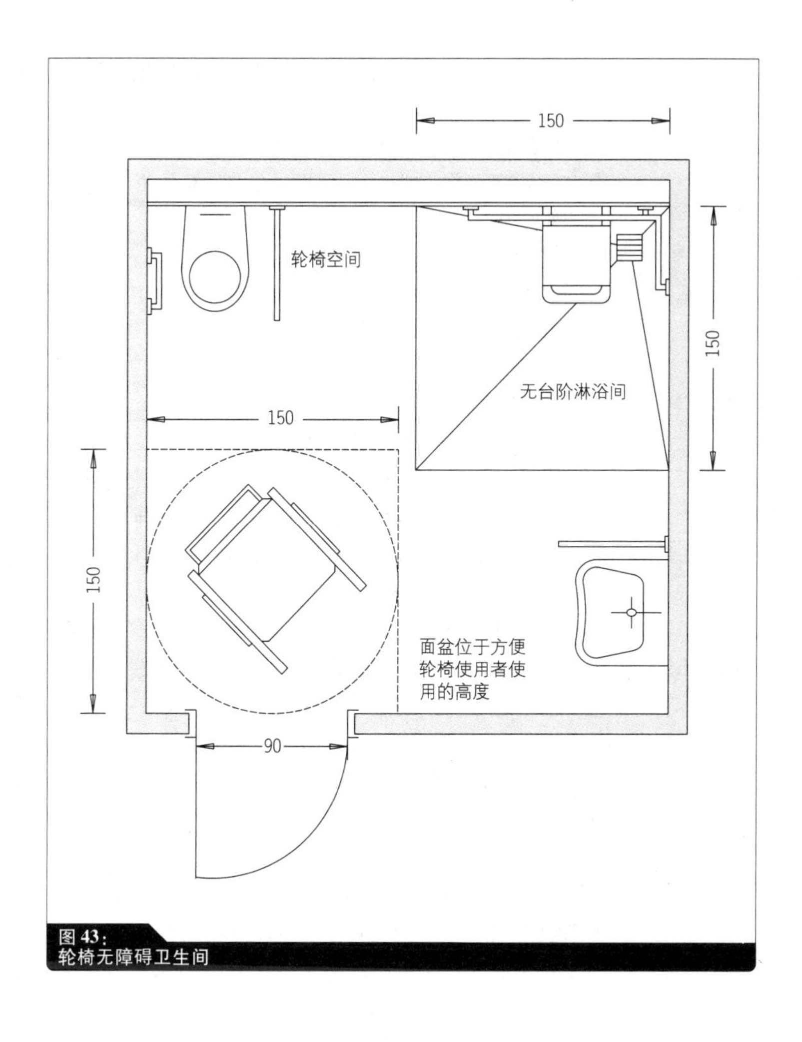

图 43：
轮椅无障碍卫生间

理工作一般在公共废水处理厂进行。

从厕所和洗碗槽中流出的废水属于重污染水，其中含有粪便和腐败物质，这种废水也被称为“黑水”。从面盆、淋浴头以及浴缸中流出的废水污染相对较轻，其污染物质仅有黑水的三分之一左右，这种废水称为“灰水”。对于废水处理厂而言，处理黑水与灰水并没有差别，因为废水处理厂的处理对象是黑水。但对于自然废水处理而言，黑水与灰水之间存在非常重要的差别，因为有些自然废水处理设施只能处理灰水。

通过近些年来人们对大自然环境关注的增加，人们将关注的重点转移到了如何保护地下水、湖水以及河水的纯净。由于生物在污染超过一定的程度之后将失去其自洁能力，所以对于重污染的工业和生活废水的处理净化方法对于预防生态破坏至关重要。〉见“废水”一章中“废水处理方法”一节

但是，在进行废水处理工作之前，使用后的冷、热水必须从建筑内部的卫生间等房间排放到公共下水道系统中。这个排放的过程将在下面的部分进行详细描述。

P54

建筑内部排水系统

废水管道的尺寸较饮用水管道大出很多，其功能是将建筑中的雨水和污水传递到下水道中。废水管道一般设计成具有不同管径的枝状管网形式，以确保卫生设施能够持续无碍地排放污水。建筑中的废水一般利用重力进行排放，故废水管道一般设计成竖直的或者具有不小于2%的坡度，将建筑中的废水从建筑下方和外部排出。非常重要的一点是需确保排出的废水不会回流到建筑内部。

溢流水位

溢流水位指的是下水道中的水位上升到一定水位之后，废水将流向下水管道系统中的某一特殊位置。〉见图44一般来说，在没有当地相关部门的特殊规定时，可以将路面或者路缘石连接处的上表面作为溢流水位。溢流水位是在溢流发生的情况下可以达到的最高水位，因此废水无法达到建筑内部更高的地方。在雨季的时候最可能发生溢流，此时合流制排水系统的风险最大，因为该系统需要同时排放雨水和废水。〉见“废水”一章中“废水处理方法”一节当然，溢流也可能在分离式排水系统中出现，比如当管道发生阻塞的时候。

处于溢流水位以下的地下室中并与下水管道相连的卫生设施，一直都存在下水道中的水回流到建筑中并造成损害的危险。故在所有的连接位置均需要设置防溢阀或者废水提升装置。〉见“建筑内部排水系统”中“防护措施”一节

实例：

德国一个四口之家将向下水道中排放大约100kg洗涤剂。虽然有利于环境保护的洗涤剂的出现能够降低污染负荷，但也只是对污染问题进行了非常微小的改善。

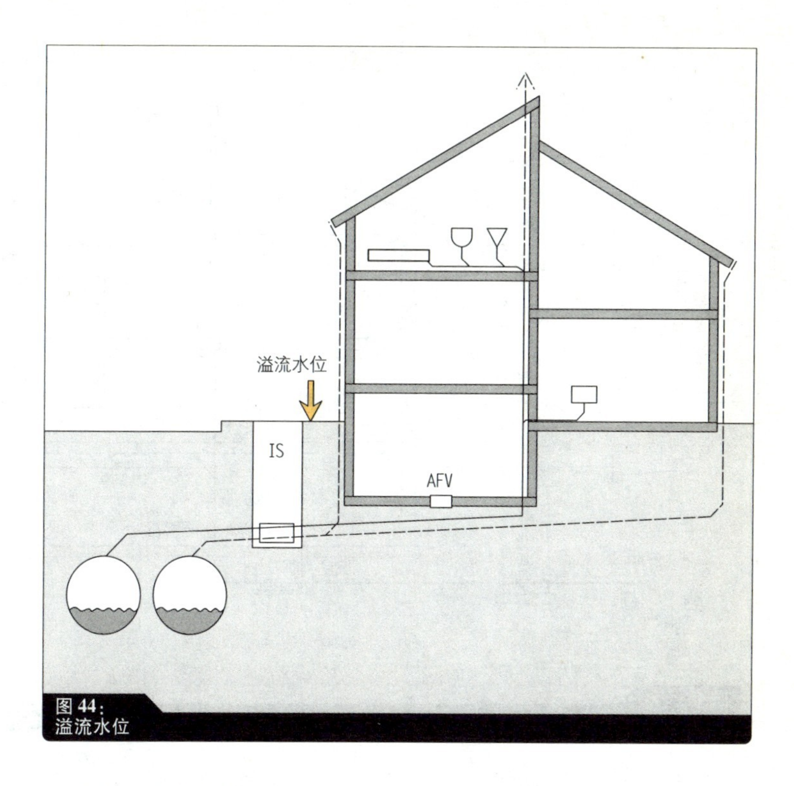

图 44：
溢流水位

P55
单独废水管及共用废水管

系统组成及管道

排水系统由若干不同的部分组成，它们相互连接、共同工作，将废水排放到公共下水系统中。〉见图 45 单独废水管与各个卫生设备单独相连，然后与共用废水管相连，而共用废水管则将卫生间中的所有不同废水来源合到一处。〉见图 46 共用废水管具有 2% 的坡度，并用最短的方式与竖向废水总管相连，然后总管用尽可能竖直的方式将废水从等直径的管道中向下排放。水平布置的废水管必须具有一定的坡度，

提示：

当废水发生溢流时，公共下水道中的废水将有可能进入建筑内部与之相连的管道系统中。根据连通器原理：当相互连接的导管上部敞开时，不论导管的形状如何，所有导管中的液体水位高度相同。

提示：

欧洲规范（EN 12056）针对的是建筑内部的重力排水系统，而 EN 752 则是针对建筑外部的排水系统。两本规范都只是给定了框架条文，具体实施需要结合各国国家规范条文，而且也允许不同地区的条文要求之间存在一定的差异。

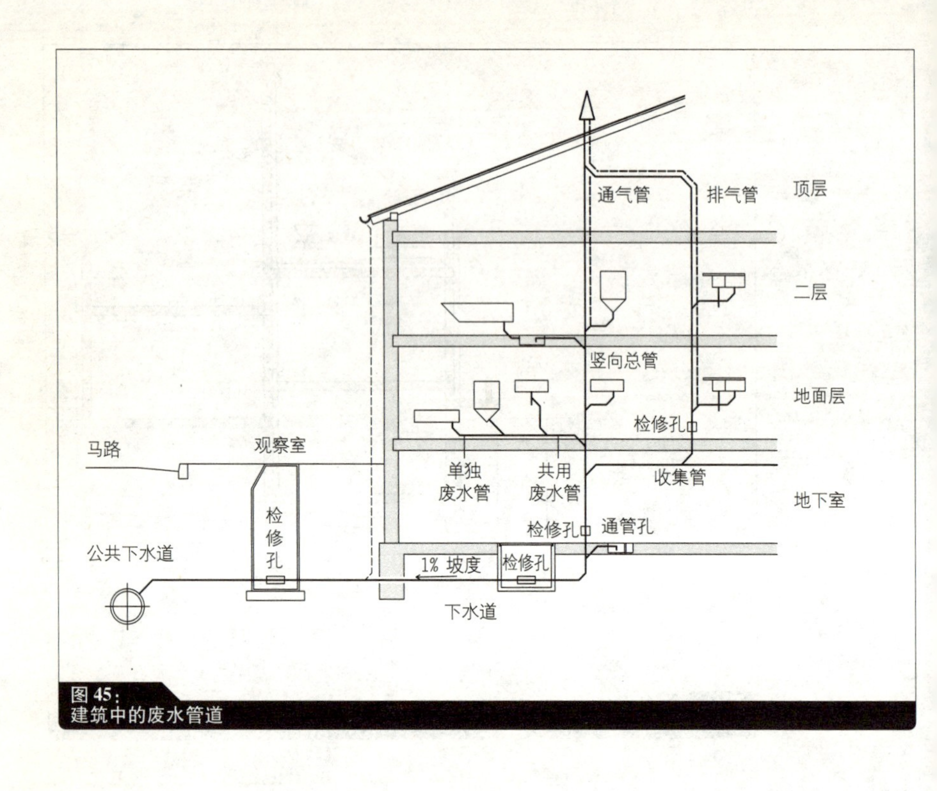

图 45：
建筑中的废水管道

以确保废水流过之后不会在管道中留下残余物。一般管道均采用与排水方向成45°夹角的方式与排水管相连，以防止废水在连接处聚集。相邻的普通排水管道应该交错间隔与竖向总管相连，以避免从排水管中排出的废水进入相邻的排水管道。对于单独废水系统，排水管之间可以采用螺丝、熔焊或者快速接头进行连接。

竖管，下水道及通气管

竖管指的是各层楼板以下与下水道相连的竖向排水管道，而下水道则是将废水排放到公共下水道或者污水沟中。竖管必须设置通气装置，以避免污水发生回吸进入卫生设施中。当多个设置突然同时使用竖管排放污水时，就可能因为竖管中产生的压力差造成污水的回吸。当竖管的总长度超过4m时，竖管将穿过不止一层建筑，此时必须在顶层楼板设置通气管通向屋顶之外，且通气管的直径不得小于竖管直径。或者采用在屋顶下面设置特殊排气阀的方式取代伸出屋面部分的通气管。如果是采用伸出屋面的通气管的形式，则要求排气口距离屋顶窗或者老虎窗的距离不得小于2m，或者设置在这些设施上面不小于1m

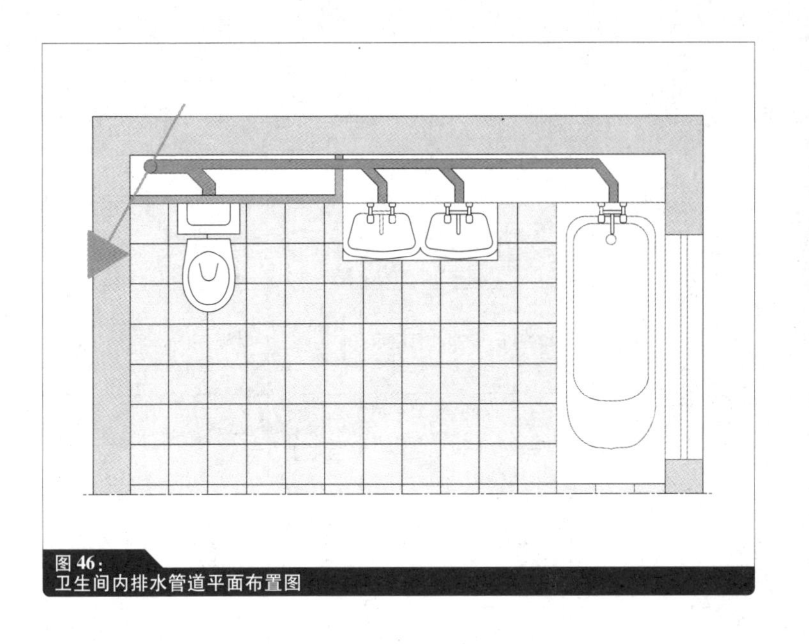
图 46：
卫生间内排水管道平面布置图

的位置，以避免废水管道中难闻的气味通过窗户进入建筑内部。

设置于竖管下方的下水道必须埋入地下足够的深度，以达到冻土层以下。下水道将建筑内部的废水排放到与之相连的室外下水道中，然后排放到公共下水道中。当建筑设有地下室，而且公共排水管道所在位置较高时，显然无法采用普通的浅埋下水道的方式进行室内废水的排放。此时可以采用在地下室顶板下方设置收集管的替代方式。

疏通通道

对于埋管和收集管，只有节点处连接管角度不小于 45°时，才能确保废水的顺利排放。另外，通管孔以及类似设施的间隔距离最多不得大于 20m，以确保管道在发生堵塞的时候可以便利地进行疏通。对于竖向总管，在管道的最低处必须设有疏通孔，因为竖向总管一般最早在此处发生堵塞。

信息标识牌

信息标识牌用来指明地下公共排水系统的位置。标识牌通常设置在附近建筑的墙体或者标志杆上，标识牌上的数字表示的是相连下水道的方位和距离。

制图符号

为了增加图形可读性，同时标明与之相连的卫生设施的布置和数量情况，在排水系统设计图中使用了很多不同的制图符号。〉见图 47 与“给水”一章中所述的内容相似，在排水系统设计图中要求采用适当的制图符号，标明与排水管道相连的卫生设施的正确位置。在剖面图

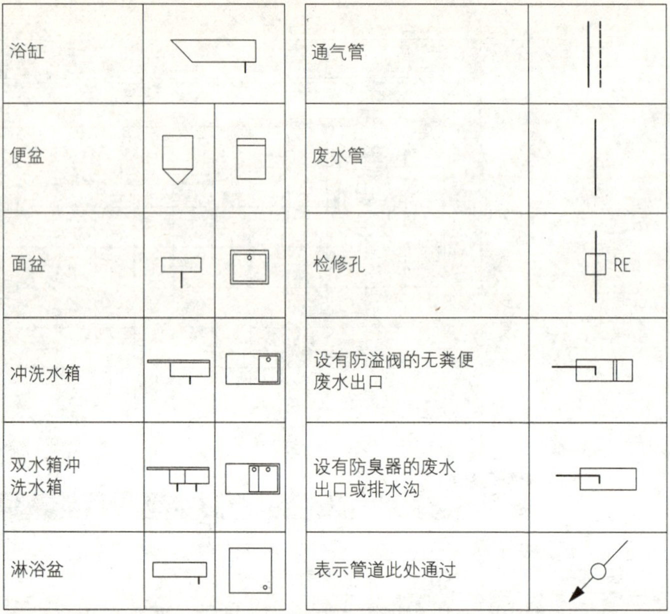

浴缸			通气管	
便盆			废水管	
面盆			检修孔	RE
冲洗水箱			设有防溢阀的无粪便废水出口	
双水箱冲洗水箱			设有防臭器的废水出口或排水沟	
淋浴盆			表示管道此处通过	

图 47：
常用制图符号及解释

中，管道系统中排水管道布置和卫生设施均采用了剖面图形的表示方法。在剖面图中，面盆、淋浴头、浴缸以及坐便器等看上去均相邻布置，并与一单独的普通排水管道相连。〉见图 45 排水管道采用前文所述的 45°布置方式，能够正确地表示出水流的方向。

P58

管道尺寸计算

废水管道的尺寸取决于与之相连的卫生设备的数量、类型以及用水量，而卫生设备的选用则取决于建筑的舒适和便利程度需求。每一个卫生设备都有各自所对应的废水量连接值（DU）以及所需废水管的最小截面尺寸。〉见表 6 在计算排水系统废水管道所需的名义尺寸时，最重要的依据是预估废水流量 Q_{ww}（其单位为 L/s）。排水指标 K 用来衡量卫生设备产生废水的频率。因此，不同建筑会因为其类型和使用功能不同，其排水系统也设计得大不相同。比如，校园内和公共场所的卫生间的使用频率要远远高于住宅中卫生间的使用频率。

表 6：
单独废水管的常用废水量连接值和所需管道尺寸

废水产生设备	废水量连接值（*DU*）	管道直径
面盆	0.5	DN 40
排水口可封堵的淋浴间	0.8	DN 50
排水口无封堵的淋浴间	0.6	DN 50
浴缸	0.8	DN 50
容积为 6L 的坐便器水箱	2.0	DN 100
拥挤为 4～5L 的坐便器水箱	1.8	DN 80～DN 100

所有废水产生设备的废水量连接值进行总和，用来计算共用废水管道、竖向总管以及下水道管道的尺寸。所以，下水道的尺寸需要根据与之相连的所有废水产生设备的废水量连接值的总和进行计算。打个比方，一般来说，与坐便器相连的水平废水管道所需的最小尺寸不得小于 DN 100，也就是说废水管道的内径不得小于 100mm。故竖向总管的尺寸也不得小于 DN 100，对于有多个连接设备的排水管道，其废水量连接值可以按照以下公式进行计算：

$$Q_{ww} = K \times \sqrt{\sum(DU)} \quad （单位：L/s）$$

其中，Q_{ww} 为废水总量；

DU 为设计单位（连接值）；

K 为无量纲排水指标，表示的是废水产生设备的使用频率（住宅取 0.5；学校、饭店、酒店取 0.7；公共建筑中常用设备取 1.0）。

P59

管道材料

废水管道可以选用陶土管、铸铁管、钢管、纤维水泥管或者塑料管等，但雨水落水管可能采用铅管。陶土管由于能够承受较大的荷载，故一般用在埋入式下水道中。铸铁管和纤维水泥管则可以用在所有建筑和地面的下水管道中。由于它们的密度较高，故在减小废水流动噪声方面具有很大的优势。若废水中具有腐蚀性物质，则一般选用钢管和不锈钢管，比如实验室中的废水管道。塑料管道则具有经济性最佳的优点，由于其具有重量轻、耐腐蚀等优点，大多用在住宅建筑

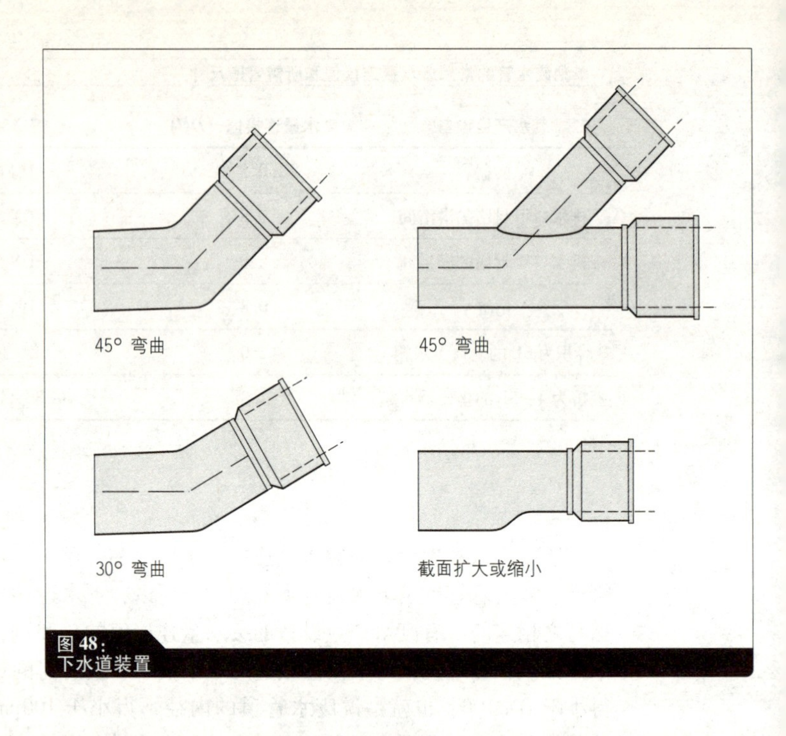

图 48：
下水道装置

的废水管道中，但部分高质量的塑料管道也在工业建筑、商业建筑的废水管道中有所应用。要求所有的塑料构件都具有防火的性能。

所有不同材料的废水管道均是按照较短尺寸的标准件进行制作，然后在施工时相互之间采用快插式套筒、螺纹或者密封方式进行连接。对于雨水落水管，还可以采用折边或者焊接的方式进行连接。通过连接和加工，可以将废水管道加工成弯曲、分支、扩口等多种不同的形式。〉见图 48

P60

防臭器

防护措施

防臭器的主要目的是防止难闻的味道从废水管中散发到室内，一般设置在每个卫生设备的出水口下方。防臭器的方式多样，但它们的工作原理大都与 U 形弯防臭器相似。U 形弯防臭器的使用最为广泛，具有很好的流动性能。U 形弯防臭器由一段直径不小于 30 ~ 45mm 的弯管构成，弯管部分始终有部分水存留 〉见图 49 左图而底部防臭器比较容易发生堵塞，故使用得并不多。〉见图 49 右图两种防臭器中所存留的水起到液封作用，防止废水管中的味道散发到室内。

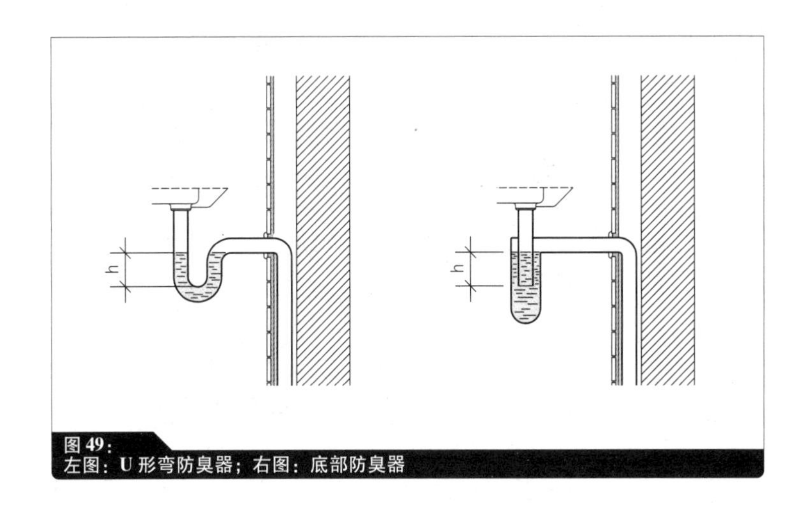

图49：
左图：U形弯防臭器；右图：底部防臭器

地面排水口

当浴室中设有淋浴间或者洗衣机时，设置地面排水口是一个比较好的选择。地面排水口主要用在公共建筑和游泳池中，其使用的材料包括铸铁、不锈钢、铜以及塑料等。地面排水口同样可以设置在楼板结构中，但要求楼板具有所需的最小安装厚度。设有地面排水口的地面在各个方面应具有不小于1.5%的坡度（朝向排水口），同时需要进行防水隔离。〉见图50 由于地面防水口一般设在房间的中间位置，往往在将排水管道与竖向总管连接时存在一定的困难。地面出水口一般与DN 50或者DN 70的废水管相连，此时要求废水管道具有不小于2%的坡度。

防溢阀

在前文讨论“溢流水位”的内容中已经指出：对于设置在溢流水位以下的废水排除设备均需要设置密封的防溢阀，以防止污水回流进入建筑内部。发生污水回流的可能性是存在的，特别对于公共下水道系统。当暴雨发生的时候可能造成公共下水道系统超负荷，当公共下水道中的水位过高时，就可能导致废水回流到建筑内部的卫生设备中。〉见图51

防溢阀通常是由一个马达驱动的止水阀和气压闸阀或者是自动、手动截止阀（紧急关闭）组成。然而，处于溢流水位以上的楼层中的所有废水管道禁止通过防溢阀进行废水排放，而是应该连接在防溢阀的下方，否则建筑中可能会有内部排放的废水溢出的危险。

废水提升装置

对于设置在溢流水位以下的废水排放设备，如果由于位置太低而无法满足足够的排水坡度将废水排放到公共下水道中，则需要利用废

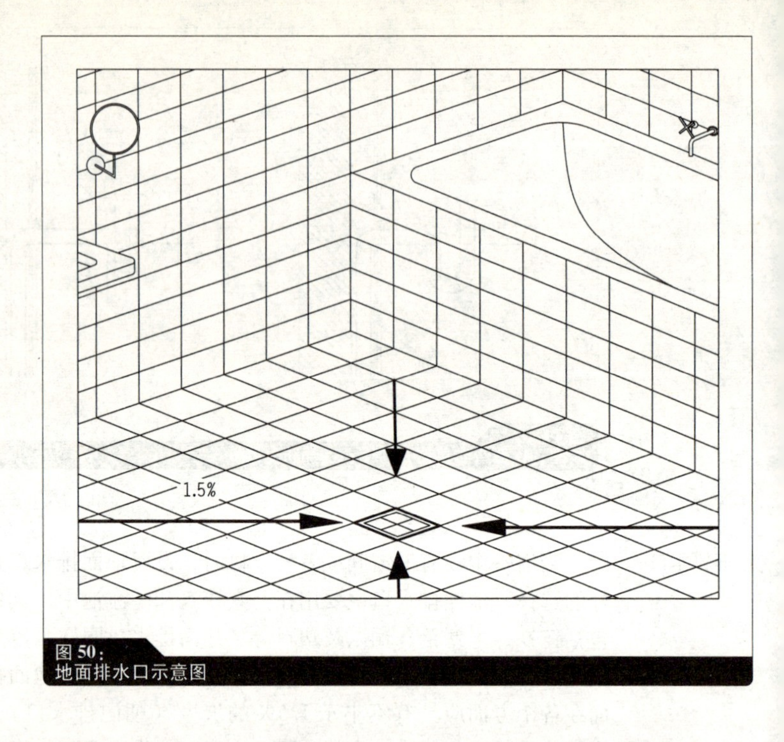

图 50:
地面排水口示意图

水提升装置进行废水排放。废水提升装置将收集的废水（可能为黑水或者灰水）通过水泵和压力管道提升，并经过防回流立管排放到公共下水道系统中，其中防回流立管的最高点应在溢流水位之上，〉见图 52以防止废水回流到建筑内部。立管之后是与公共下水道相连的普通的具有排水坡度的废水管道。

P62

废水处理方法

分离式下水道系统和共用下水道系统

废水将排放到单独下水道系统或者共用下水道系统中。对于共用下水道系统，生活废水和工业废水连同雨水一起排放到下水道系统中；对于分离式下水道系统，雨水直接流向开放的水体或者水道中（常见的如排水口等），只有受到污染的水才排放到公共下水道中。〉见图 53 目前，对于建筑所在区域以及在进行废水管道设计时，都将雨水和废水进行分别排放；甚至对于路面下方的共用下水道系统，许多国家也计划在不久的将来改造成为分离式下水道系统。

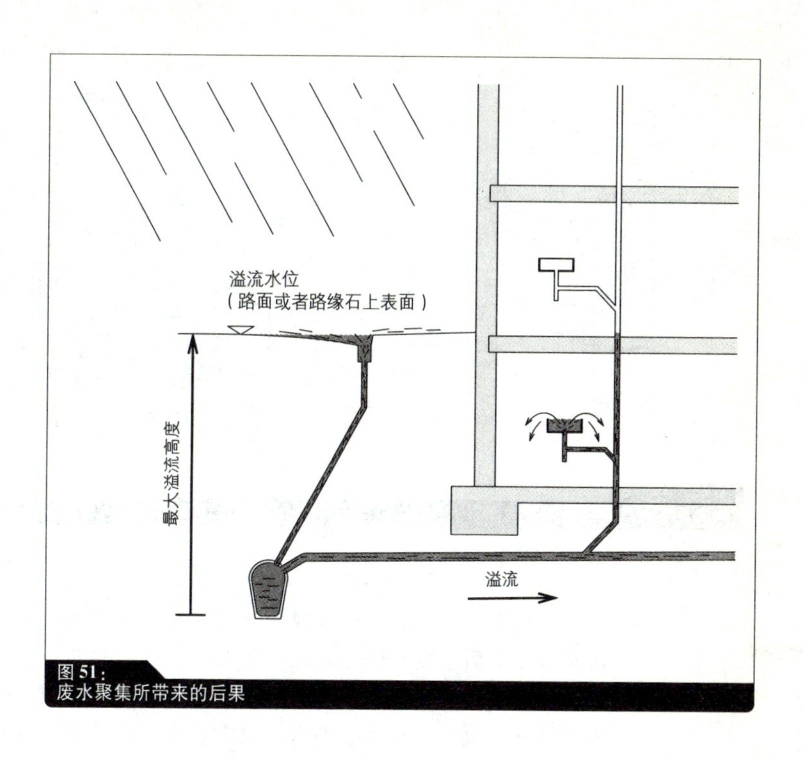

图 51：
废水聚集所带来的后果

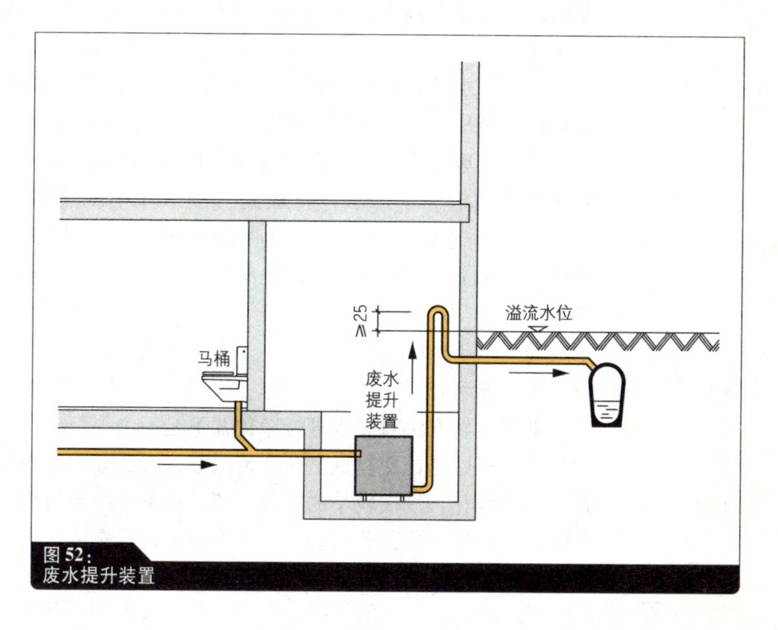

图 52：
废水提升装置

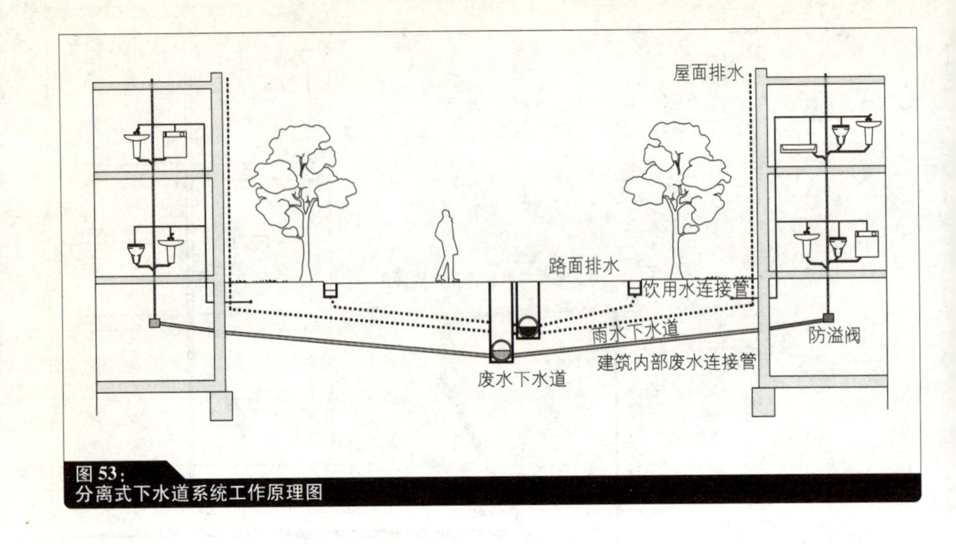

图 53：
分离式下水道系统工作原理图

进行上述改造的原因是因为当降水在公共下水道系统中与普通污水混合之后会变成脏水，而导致废水量急剧增加。如果采用分离式下水道系统，不仅会减小下水道的排水压力，同时还会大大降低废水处理的成本。从减小废水排放量和维持自然地下水位两个方面考虑，应该确保雨水尽可能地排放在建筑的宅地周围或者直接渗透到土地中。

废水的净化

市政废水处理厂在进行废水净化的时候，首先去除的是废水中的大颗粒杂质，然后采用生物方法去除废水中的细菌，最后采用化学方法除去废水中的磷酸盐、重金属以及硝酸盐等，〉见图 54 在净化完成之后通过废水排水口排放到自然水体中。然而，即使我们对废水进行了过程复杂、耗资昂贵的净化处理，其中所含的过多的植物营养和污染物质仍然将进入自然水体中，从而刺激植被的过快生长。

P65

天然废水处理系统

采用天然方法对废水进行净化并不是一个很新的方法，这种方法的另一个好处是花费较低。通常情况下，天然废水处理设施针对的都是一些本地化、规模较小的排放系统。在偏远的乡村地区，由于距离公共排水设施太远，若与公共排水设施连接排水成本过大，故采用天然方法进行废水处理，而且已经有了数年的历史。

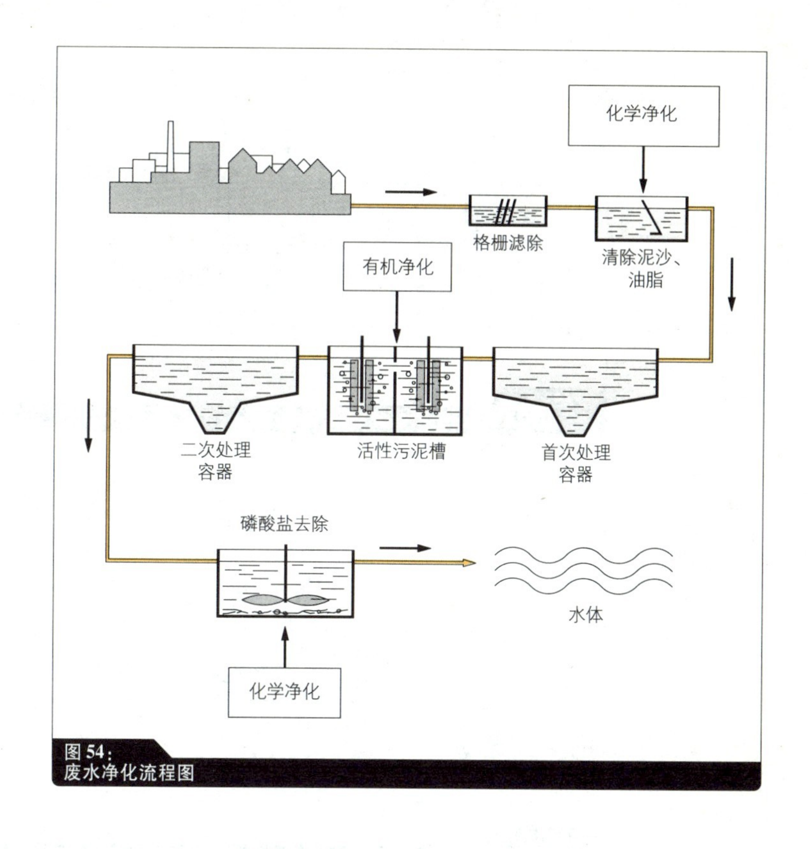

图 54：
废水净化流程图

近些年来，由于废水处理的成本以及水质问题日渐突出，一些重视环境保护的设计者重新考虑了分散化的天然废水处理方法。大量的生态住宅开发选址在芦苇湿地废水处理系统附近，使得住宅所产生的废水可以在建筑所在地内得到净化处理。这种做法大大缓解了公共废水系统的压力，同时也增强了人们对于自然界水循环的生态观念，并将保护自然界水循环的责任落到了个人的身上。

天然废水处理方法不需要太高的能耗，也不需要传统废水处理厂耗资巨大的成本，但却能够收到很好的废水净化效果。整个废水净化过程几乎不需要消耗外界的能量，但需要占据较大范围的土地。天然废水处理系统的设计主要应该考虑的是将要净化废水的污染程度。

芦苇湿地废水处理系统

芦苇湿地废水处理系统是天然废水处理最常见的方法：一般建在废水池的周围，废水池中种植芦苇，其净化功能主要来自于动、植物

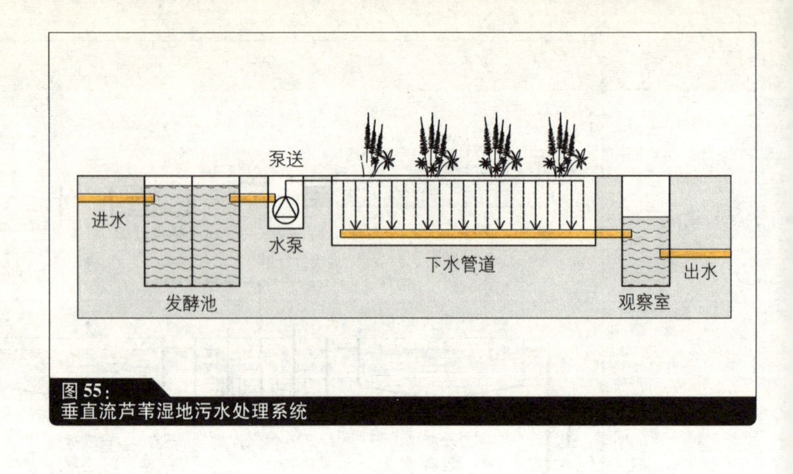

图 55：
垂直流芦苇湿地污水处理系统

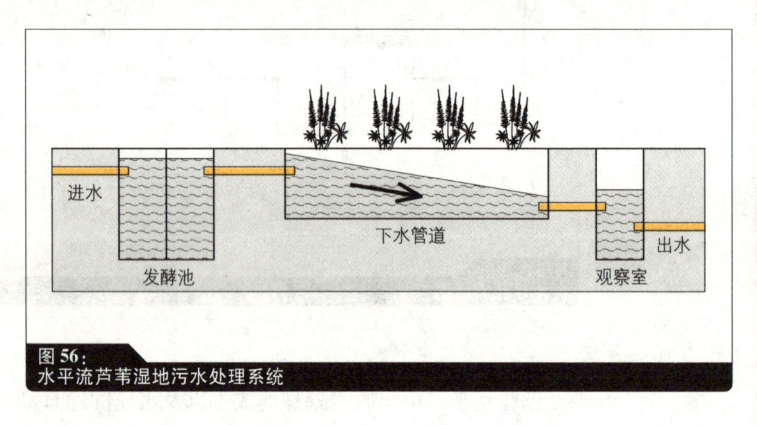

图 56：
水平流芦苇湿地污水处理系统

微生物。所以说，净化废水的并不是芦苇本身，而是生长在其根部的微生物。这些微生物能够消耗废水中的营养物质，从而起到净化废水的作用。废水池的池底一般是砂滤层，废水采用垂直或者水平的流动方式。在上游位置一般设有一个发酵池或者三室化粪池，用来除去废水中的固体物质，但必须使空气能在其中连续流动，以确保发酵效果。

首先，淋浴间和面盆中排出的灰水以及厕所中排出的黑水从一个单独的排水管道中利用重力排出建筑外部，然后流到发酵池中，而发酵池的主要功能是排除废水中的大颗粒杂质。接下来，采用水泵将废水从发酵池中泵送到芦苇区。若采用垂直流处理系统，则占用的区域较小，但是过滤层埋设的深度较深。〉见图 55

图 57：
储水池

另外，若采用水平流处理系统，则所需的深度较小，但占用的区域较大。污水将从芦苇区缓慢地流过，并在此过程中得以净化。〉见图56 处理系统类型的选择主要取决于建筑外部的可用区域大小，也有一些污水处理方法为了达到更好的处理效果兼用了以上两种处理方式。

芦苇区实际上是一个简单的砂砾过滤层。由于大部分的废水从过滤层渗入土壤中，所以芦苇区看来并不像开放的水域，而更像是陆地。废水在经过多级过滤之后将来到观察室，在此处将对处理后的废水水质进行规律的检测。如果检测水质满足要求，则可以从出水口直接排放，或者排放到储水池中作为休闲景观。〉见图 57 另外，经过处理后的水也可以作为厕所冲洗用水。〉见“废水”一章中“废水利用”一节

天然废水处理方法需要占用较大的区域才能有效地发挥功能，达到很好的废水处理效果，当进行黑水处理的时候更是如此。但是，这种方法具有天然和美观的优点。所以说，和传统废水处理厂相比，更加提倡采用芦苇湿地废水处理方法，而且这种方法不能简单看成是传统废水处理方法的生态改进方法。对于远离公共下水道系统的建筑，采用天然废水处理方法可以很好地对所有建筑内废水进行有效的净化。

P68

雨水排放

雨水也是建筑内水循环的一部分。雨水从屋顶流出，对于建筑立

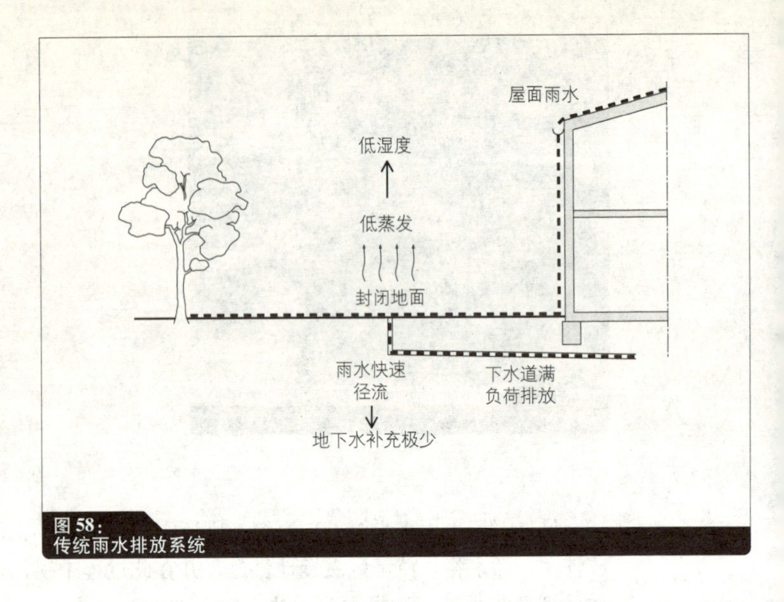

图 58：
传统雨水排放系统

面而言，雨水是影响洁净的脏水，所以必须和生活废水一样进行排放。

对于自然水循环而言，理想的情况是雨水渗入土壤中成为地下水。而城市中的实体路面对雨水直接渗入土壤中形成地下水造成了阻碍，取而代之的是利用下水道系统进行排水。〉见图 58 在降雨量较大的时候，下水道的排水能力往往无法满足需要，导致大量的未处理污水与雨水一同流入河流和湖泊中。

对于雨水排放，如今的方法并不是要用最快的速度将雨水排放到下水道中，而是采用更加精确的方法减慢雨水流动速度，避免雨水和污水混合。在设计和选择雨水排放系统类型的时候，设计者需要考虑所在地区的降雨量、降雨频率、地表类型以及地下水位高度等多方面的因素。

P69 雨水渗透

为了维持自然界水循环，对于建筑区以外（特别是住宅区）的空地、小路以及广场等应该尽可能设计成雨水可以渗透的地面，比如草地或者砂砾等。对于建筑区以内的地面，也可以采用相应的措施来增加地面的渗透性，促进雨水能够以一种自然的方式进入土壤中，从而提高地下水位。

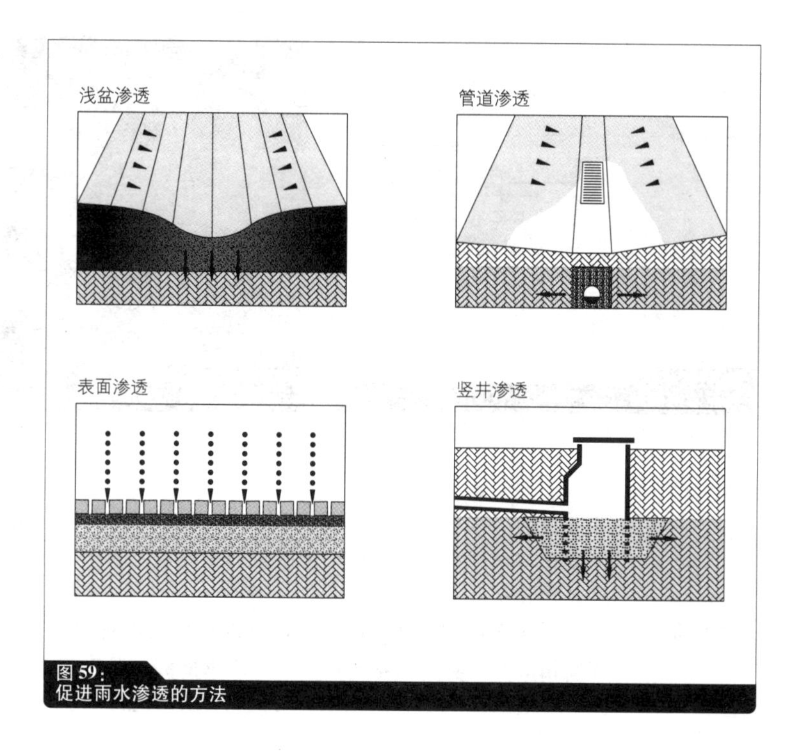

图 59：
促进雨水渗透的方法

影响雨水渗透最主要的因素是地面性质。土壤的砂质越强，地面的可渗透性越强，雨水渗透即可毫无困难地自然进行。如果土壤的壤质或者黏质太强，阻碍了雨水的渗透，则需要采取特殊的措施来帮助雨水渗透。比如，采用填草浅盆、管道、浅坑等渗水措施。〉见图 59 这些措施能够减缓雨水的排放，在雨水较大的时候能够将雨水留在土壤中，防止下水道水位过高和超负荷工作。除此之外，这些措施所保留的雨水能够极大地改善城市的气候。

在下水道旁设置渗水措施还能够很好地节约成本，但对于一些更加复杂的渗水系统，比如设有雨水池的绿色屋顶、大规模的渗水系统（比如填草浅盆等），由于维护费用较高，也可能造成成本增加。

表面渗透

表面渗透指的是雨水不在地面暂时聚集而直接渗透到土壤中。蜂窝型草坪路面是表面渗透的一种形式；〉见图 60 和图 61 另一种表面渗透形式是采用渗水路面砖。这两种形式都适合用在停车场、花园和使用较少的路面中。草坪和砂砾道路也能够增加下层土壤的渗水量。雨水

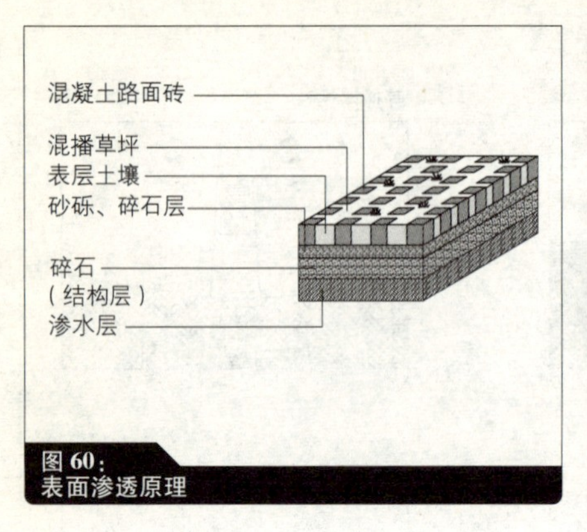

图 60：
表面渗透原理

图 61：
利于表面渗透的蜂窝草坪路面

在经过表层渗透之后（即使是经过了表层土壤之后）能够得到初步的净化，自雨水从更深层的土壤直到渗入地下水中，土壤一直能够发挥其净化功能。

浅盆渗透

浅盆渗透和洼地渗透是一种可以延缓雨水排放的表面渗透形式。浅盆指的是一块种有草坪的凹地，雨水能够在其中聚集若干小时。〉见图 62 在这段时间里，凹地里的雨水能够缓慢地渗入土壤中，并最终进入地下水体里。

浅盆渗透相比表面渗透所需的场地较小。用浅盆深度 30cm 进行估算，所需的浅盆面积大约为其对应排水屋面面积的 10% ~20% 。在渗入不同土层的过程中，雨水也得到了净化。渗透浅盆的施工成本较低，而且不需要太多的维护费用。另外，浅盆还可以作为设计元素，与休闲设备、绿化场所设计成一个整体。

管道渗透

管道渗透同时采用了两种渗透方式：表面是浅盆渗透，下面是设置在砂砾层上的排水管道。浅盆表面大约 30cm 厚度，既可以聚集雨水也可以过滤雨水。聚集的雨水通过某一点或者某个区域进入排水槽中。排水槽中填有砂砾层，并铺设有无纺布过滤层。多余的雨水经过排水管道缓慢地流向排水口，或者排放到公共下水道系统中。在缓慢排放的过程中，大部分的雨水从多孔的排水管道中再次渗透到土壤中，故经过排水口排出的雨水只占很少的一部分。在路面排水条件较差且雨水量较大的情况下，适合采用管道渗透的方式。

P71

雨水保留

在大型城镇和城市里，可以通过雨水的保留减少数百万升的废

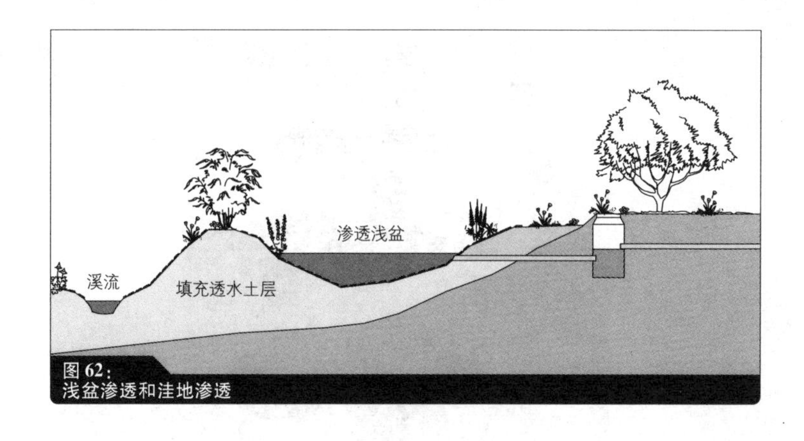

图 62：
浅盆渗透和洼地渗透

水。雨水保留系统的主要功能是延缓和减少雨水直接流向下水道系统中。大多数情况下，雨水通过绿色屋顶和雨水池进行保留。绿色屋顶保留雨水的能力在一定程度上取决于屋面基底材料，通过聚集雨水，绿色屋顶可以将排放到下水道系统中的雨水减少三分之二。绿色屋顶能够改善城市气候，特别是局部区域的微气候，从其中蒸发的雨水能够降低夏日的高温和抑制空气中的灰尘。

绿色屋顶

绿色屋顶可以是粗放型或者密集型。粗放型绿色屋顶的基底材料厚度在 3 ~ 15cm 之间，而密集型绿色屋顶的基底材料厚度在 15 ~ 45cm 之间。两种绿色屋顶均在传统屋顶的形式之上增加了一个特殊的屋面层，以防止植物根的生长破坏屋顶结构。在该屋面层之上是排水层，用来排放积存的雨水。然后上面才是植被层。〉见图 63 屋面结构可以设计成为保温屋顶、非保温屋顶或者倒置屋顶等多种形式，其主

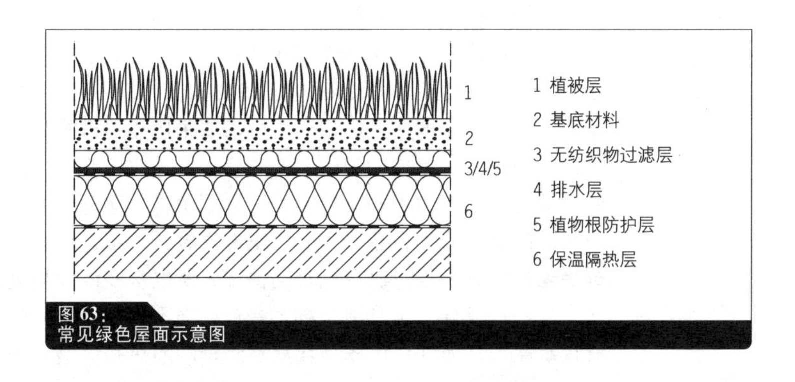

图 63：
常见绿色屋面示意图

图 64：
将雨水排放作为特色标志进行设计

要的差异在于屋面保温层和其中通风层的位置。

粗放型绿色屋顶的基底材料厚度一般在 3 ~ 7cm 之间，适合种植苔藓类和多汁植物类等对雨水和营养要求较低的植物。对于草地或者较多叶的植物，其需要的雨水稍多一些，则需要更深一些的土壤层。对于种植有生长较快的草地、多年生植物或者木本植物的密集型绿色屋顶，若基底材料的厚度约厚，则屋面保留雨水和蒸发雨水的能力越强。然而，其种植成本越高，同时屋顶荷载也成比例增加。

对于平屋顶，如果屋顶结构满足密闭性、舒适性的要求，同时也可以承受屋顶基地材料所带来的附加荷载，就可以设计成为绿色屋顶。只要其屋顶倾斜角度不大于 15°，就不需要另外设置相关的安全措施。但对于坡度更大的屋面，则需要采用相关措施预防屋顶的泥土滑出屋檐外。

绿色屋顶的植被和土壤层可以对雨水进行净化。所以说，可以将净化后的雨水收集在容器中，用来冲洗厕所。不过，一般收集的雨水都会全部用完，很少出现剩余的情况，所以从经济性角度出发，没有必要另外再设计一个管道系统。绿色屋面不仅仅能够减轻下水道系统的排水压力，另外对于屋顶在夏、冬季的隔热和保温方面能够起到非常大的帮助作用。

储水池

储水池一般在底部设有密封层，与渗透盆的不同之处在于储水池中的水一直保留，并不渗透到下层土壤中。储水池可以当做真实的栖息地进行设计，其周边可以进行绿化，能够为很多野生动物提供活动和栖息场所。屋面的雨水主要通过多个很小的水流汇集到储水池中，当雨水较大的时候，储水池中的雨水溢出，流向周边的渗透浅盆中。住宅开发设计时，在公共开放区域可以将储水池作为一种特色标志进行设计。

雨水设计

在对住宅区的空地和休闲公园进行设计的时候，可以将雨水渗透浅盆和雨水池作为特色标志融入其中。不再采用地下下水道进行排水，而是采用开放的小河道、小溪流等，这种方式能够使住户在休闲时心情更加愉悦。〉见图 64

P74

废水利用

由于污染问题的加重，污水处理的成本越来越高，而且处理流程也越来越复杂。另外，实际上只有很少几个地方需要使用很高质量的饮用水。但让人很难理解的是，有大量的雨水和轻微污染的污水却未加使用而直接排到了下水道中。近些年来，出现了越来越多的将雨水和灰水置换成饮用水的方法。

P74

雨水利用

雨水利用可以节约饮用水、减轻下水道排放和污水处理厂的压力。只要雨水中没有重金属和有毒物质，就可以安全地用在厕所冲洗、花园浇灌以及洗涤中。雨水的水质取决于雨水降落所在的地点以及其流过表面的材料性质。打个比方：屋面可能会有灰尘和鸟类粪便，故其上富有微生物。另外，从高速公路和停车场的下水道中排出的污水由于可能被汽油残余物污染，所以不适合再次利用。在这种情况下，则建议放弃对雨水的利用。在进行建筑物管道设计的时候，应该尽可能将一些不太复杂的雨水利用系统包含其中。〉见图 65 和图 66

提示：

关于户外雨水排放设计的建议可以参考本套丛书中的《水景设计》一书，中国建筑工业出版社预计于 2011 年 6 月出版。

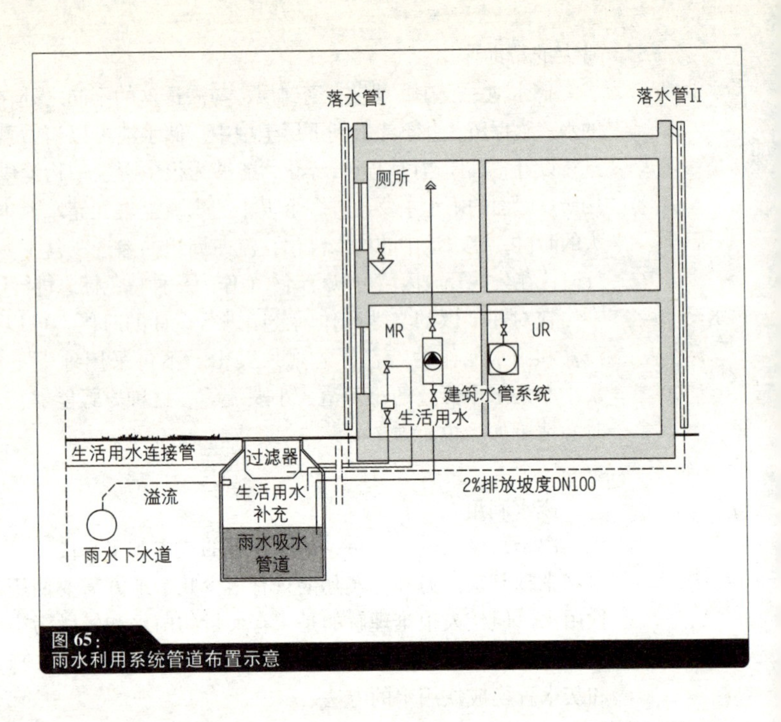

图 65：
雨水利用系统管道布置示意

屋面雨水收集面积

屋面尺寸和表面性质对雨水的收集和利用至关重要。如果屋面由光滑材料覆盖，则大部分雨水可以沿屋面流出；如果屋面由多孔渗水材料覆盖，则部分雨水将被屋面吸收然后蒸发掉。一般常用的屋面材料，包括陶土瓦、混凝土屋面瓦或者石板瓦，均适合用来收集雨水。如果采用金属屋面，在雨水收集系统为洗衣机供水的时候可能会导致衣物变灰。如果仅仅是将雨水用来冲洗厕所，那么采用金属屋面将不会带来任何害处。

收集到储水箱的雨水量主要取决于降雨强度、频率以及屋面的径流系数。径流系数 0.75 指的是屋面上 75% 的降雨将经过屋面流向落水管，并收集到储水箱中。根据屋面性质的不同，径流系数在 0.0 ~ 1.0 之间，越光滑的屋面径流系数越高。〉见图 67

储水箱

储水箱指的是用来接收和储存由屋面流出并由落水管所排放雨水的容器。储水箱的规格和设置位置各有不同。如果建筑没有地下室，则建议采用埋入式储水箱。如果设有地下室，则建议采用不透明材料的储水箱，并将储水箱设置在温度较低且光线较暗的地下室中，以防止雨水中的细菌和藻类生长。储水箱的容积各有不同，最高可达

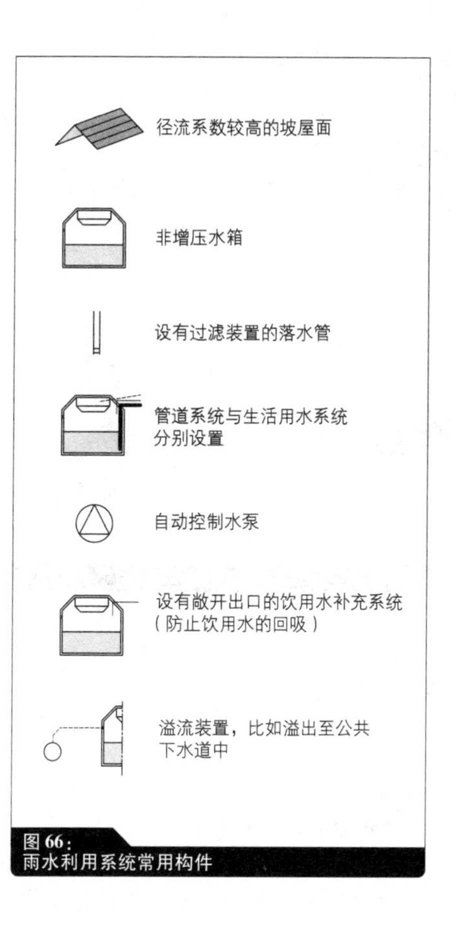

图 66：
雨水利用系统常用构件

1000L。如果储水量更大，则可以将若干储水箱连接在一起使用，或者设计一个所需任意容积的埋入式混凝土储水箱。

储水箱

在确定储水箱容积的时候，需要考虑雨水流入量和需求量两个方面的因素。可以根据气象部门提供的区域降雨图来计算可收集的雨水量。打个比方，德国不同地区的降雨量一般在每年 600 ~ 800mm 之间，而旱季的时间大致在 21 天左右。而储水箱中的雨水流入量则是根据屋面面积和屋面材料的径流系数进行计算。

✎

年雨水流入量（单位：L/a）计算：

雨水收集面积（m^2）× 径流系数（w）× 年降雨量（mm/a）

生活用水需求量（单位：L/a）计算：

每人每天生活用水需求量 × 人数 × 365 天

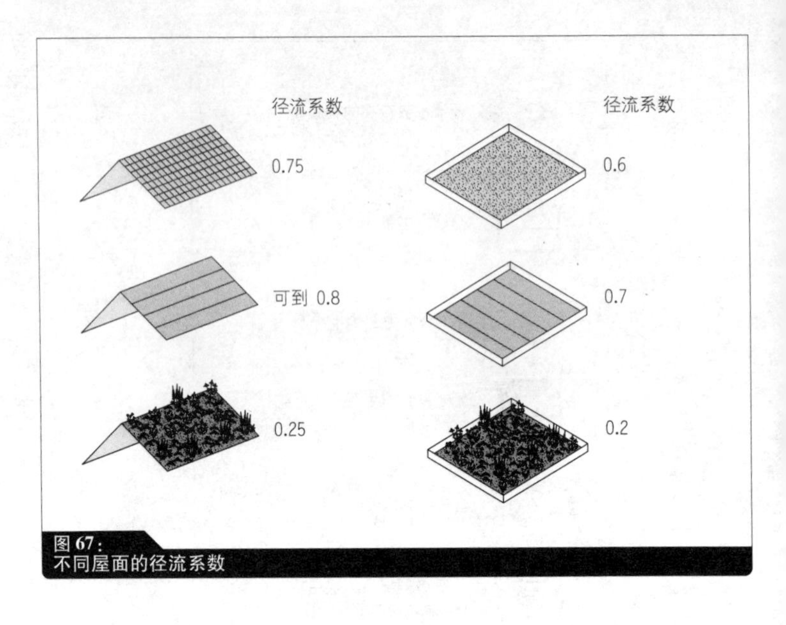

图 67：
不同屋面的径流系数

水箱储水能力计算：

$$\frac{\text{生活用水需求量}\times 21\text{ 天}}{365\text{ 天}}$$

考虑到年降雨量和生活用水需求量之间的关系，储水箱的容积设为年雨水流入量的 5% 即可满足生活需要。

饮用水补充和分配

在旱季，水箱中的水可以由公共供水系统通过露天的防冻管道或者建筑内部的生活用水供应站得以补充。生活用水供应站由一个水泵、控制装置、压力调节阀和安全设施组成，其中水泵用来传送雨水。在进行补充的时候，需要确保饮用水管道不会和雨水直接接触，以防止公共饮用水被污染。另外，必须确保水箱所无法容纳的多余雨水能够通过紧急出水口排放到公共下水道中。

注释：

在计算雨水收集量的时候，屋顶的面积是由其投影面积计算得出的，而投影的方向是从屋面向下。

图 68：
浸入式生物萃取系统

P77 灰水回收利用

在城市中心区域，水分很难渗透进城市地面，而且一般也没有足够的空间来设置芦苇湿地污水处理系统。此时，设置生物灰水处理系统可以胜任轻微污染水的净化工作。该系统一般设置在建筑的地下室中，整个系统由多个不同的具有特殊功能的构件按照污水净化流程组装而成。该系统的具体安装需要根据可利用的空间、使用人数以及预算来确定。但在使用该系统的时候，必须将黑水分离出来，排放到公共下水道中。

浸入式生物萃取系统

常见的浸入式生物萃取系统一般由一个沉淀池、一个储水池及其中的机械净化系统和一个斗轮组成。斗轮将灰水连续地传送到转动的浸入式生物萃取器中。〉见图 68 生物萃取器由一系列聚乙烯过滤管组成，并将完成灰水净化的大部分工作。过滤管在其中相当于微生物生产车间。萃取器的转动速度为每分钟 0.5 转，并始终保持一半浸在污水池中，另一半通过和空气接触为其中的微生物提供氧气。通过这个过程，萃取器中的微生物量不断增加，在适当的时候微生物将与萃取器脱离，并沉入污水池的底部。经过一系列的净化过程之后，处理后的污水可以作为生活用水使用。

过滤膜系统

过滤膜系统是进行灰水处理的另一种方法。这种方法需要首先采用通风滤网对灰水进行预处理，在此过程中通过微生物和通氧除去灰水中的有机成分。然后将污水通过若干层密集排列的过滤膜。过滤膜系统是一个密闭系统，可以比较容易地在地下室等紧凑空间中进行安装。

灰水回收利用是最环保的污水处理方法之一。该方法不仅能够更加充分地对雨水加以利用，而且能够减少使用后的饮用水经过长距离的输送进入污水处理厂，并进行成本昂贵的处理过程。同时，还可以减少污水处理量，减轻污水处理厂的工作负荷。如果处理后的污水能够渗透到土壤中，该方法还可以对地下水进行补充，并为自然界水循环提供益处。

污水处理过程或者其替代过程处于水循环过程的末端，在此之后处理过的污水返回自然水体之中。而水循环开始于饮用水的回收利用或者再供给。

结语

通过对“生活用水”的主题进行的相关分析表明：对于饮用水和污水的处理存在多种不同的方式；另外，在将来的设计中需要建筑师进行更加深入和细致的设计，将饮用水系统、排水系统以及卫生设备作为一个整体进行设计，使它们能够更有效地发挥各自的功能。

然而，如果我们希望能够在水资源利用上做到可持续发展，就要像在节能建筑中所要求的一样，还有很多的工作需要去做。虽然只有部分国家会遭受长期或者短期的干旱，但为了保证自然界水循环的长期稳定性，还需要整个人类做出全面的努力，以防止自然界水循环负担过重。与其研究出越来越复杂、成本越来越高的饮用水净化技术，还不如确保预防地下水不会被任何污染物所污染。而要想达到这个目标，不仅仅需要建筑设计师的努力，更需要采取除此之外的很多复杂的手段。

在这方面，建筑师在进行相关设计的时候可以向业主提出所有可能的建议。不仅仅是如何设计出一个美观的浴室，同时向业主建议如何可以更加节俭地使用饮用水，如何减少污水的排放。简单的措施包括采用节水龙头、尽可能多地使用雨水、将污水使用在不同用途中以及采用有利于环境保护的雨水回收利用来取代传统使用饮用水然后再进行废水净化的办法。更重要的是，这些措施能够对我们珍贵的饮用水资源起到持续的保护作用。将来，仅仅从成本和环境保护的角度出发，也将会在水资源和能源节约保护方面进行重点而深入的研究。采用太阳能进行饮用水加热能够很好地符合节约资源的理念。只要我们有这方面的意愿，那么雨水收集和灰水回收再利用的观念将会很快而且非常容易地得以实施。虽然这些措施从每一个单独来看并没有非常大的作用，但总的来看，这些措施将对水资源保护起到主要的作用，并且能够保证水循环的稳定。

P81 # 附录

P81 ## 参考文献

John Arundel: *Sewage and Industrial Effluent Treatment*, Blackwell Science, Oxford/Malden, MA, 2000

Tanja Brotrück: *Basics Roof Construction*, Birkhäuser Verlag, Basel 2007

Committee on Public Water Supply Distribution Systems, National Research Council of the National Academies: *Drinking Water Distribution Systems: Assessing and Reducing Risks*, National Academies Press, Washington, DC, 2006

Klaus Daniels: *Technology of Ecological Building*, Birkhäuser Verlag, Basel 1997

Herbert Dreiseitl, Dieter Grau (eds.): *New Waterscapes—Planning, Building and Designing with Water*, Birkhäuser Verlag, Basel 2005

Herbert Dreiseitl, Dieter Grau, Karl Ludwig (eds.): *Waterscapes—Planning, Building and Designing with Water*, Birkhäuser Verlag, Basel 2001

Gary Grant: *Green Roofs and Facades*, IHS BRE Press, Bracknell 2006

Institute of Plumbing (ed.): *Plumbing Engineering Services. Design Guide*, Institute of Plumbing, Hornchurch 2002

Margrit Kennedy, Declan Kennedy (eds.): *Designing Ecological Settlements: Ecological Planning and Building*, Cap. Water, Reimer Verlag, Berlin 1997

Heather Kinkade-Levario: *Design for Water: Rainwater Harvesting, Stormwater Catchment and Alternate Water Reuse*, New Society Publishers, Gabriola Island, BC, 2007

Axel Lohrer: *Basics Designing with Water*, Birkhäuser Verlag, Basel 2008

Frank R. Spellman: *Handbook of Water and Wastewater Treatment Plant Operations*, Lewis Publishers, Boca Raton, FL, 2003

Ruth F. Weiner, Robin A. Matthews: *Environmental Engineering*, 4th ed., Butterworth-Heinemann, Amsterdam/London 2003

Bridget Woods-Ballard et al.: *The SUDS Manual*, CIRIA, London 2007

P82 **相关规范**

EN 752	Drain and sewer systems outside buildings
EN 805	Water supply - Requirements for systems and components outside buildings
EN 806 - 2	Specifications for installations inside buildings conveying water for human consumption
EN 1717	Protection against pollution of potable water in water installations and general requirements of devices to prevent pollution by backflow
EN 12056	Gravity drainage systems inside buildings
EN 12255	Wastewater treatment plants, Part 5: Wastewater treatment plants. Lagooning process

P83 **图片来源**

P83 照片

所有的照片均由桃丽丝·哈斯－阿尔恩特拍摄。

P83 绘图

珍妮·珀汀斯（Jenny Pottins）

西蒙·卡斯讷（Simon Kassner）

海伦·韦伯（Helen Weber）

塞巴斯蒂安·巴格斯科（Sebastian Bagsik）

英迪拉·萨德利希（Indira Schädlich）

P83 **作者简介**

桃丽丝·哈斯－阿尔恩特，工学博士，德国锡根大学（University of Siegen）建筑技术和建筑生态学专业客座教授。